W9-ABN-296

Bilinear Transformation Method

This is Volume 174 in
MATHEMATICS IN SCIENCE AND ENGINEERING
A Series of Monographs and Textbooks
Edited by RICHARD BELLMAN, *University of Southern California*

The complete listing of books in this series is available from the Publisher upon request.

Bilinear Transformation Method

Yoshimasa Matsuno

Space System Designing Section
MHI Ltd.
Nagoya Aircraft Works
Nagoya, Japan

1984

ACADEMIC PRESS, INC.
(Harcourt Brace Jovanovich, Publishers)
Orlando San Diego New York London
Toronto Montreal Sydney Tokyo

ACADEMIC PRESS, INC.
Orlando, Florida 32887

United Kingdom Edition published by
ACADEMIC PRESS, INC. (LONDON) LTD.
24/28 Oval Road, London NW1 7DX

Library of Congress Cataloging in Publication Data

Matsuno, Y. (Yoshimasa)
Bilinear transformation method.

Includes bibliographical references and index.
1. Bilinear transformation method. 2. Evolution equations, Nonlinear--Numerical solutions. 3. Benjamin-Ono equations. I. Title.
QA374.M34 1984 515.3'5 84-70234
ISBN 0-12-480480-2 (alk. paper)

PRINTED IN THE UNITED STATES OF AMERICA

84 85 86 87 9 8 7 6 5 4 3 2 1

Contents

Preface

This volume may be divided into two parts. The first part (Chapter 2) is an introduction to the bilinear transformation method. This method is a powerful tool for solving a wide class of nonlinear evolution equations. In the bilinear formalism, the nonlinear evolution equations are first transformed into *bilinear equations* through dependent variable transformations. These bilinear equations are then used to construct the N-soliton solutions, the Bäcklund transformations, and an infinite number of conservation laws in a systematic way. As an example, the method is applied to the Korteweg–de Vries equation, which is typical of nonlinear evolution equations. The essential part of the bilinear transformation method may be understood by reading Chapter 2.

The second part (Chapters 3–6) is concerned with the study of the mathematical structure of the Benjamin–Ono (BO) and related equations. The bilinear transformation method is employed extensively to analyze these equations. At the same time, the relationship between the soliton theory and the algebraic equations is stressed. In Chapter 3, especially, the mathematical structure of solutions of the BO equation is clarified from the viewpoint of the theory of algebraic equations.

The materials treated in this book are current topics, including open problems. However, the contents are presented at an elementary level and are self-contained. Therefore, the maturity assumed of the reader is that of a beginning graduate student in physics or applied mathematics. Some knowledge of such mathematics as elementary partial differential equations and the theory of linear algebraic equations will be helpful for a full understanding of the materials.

Throughout the text, the discussion is restricted to the mathematical aspects of the problem, and accordingly, for the physical background, the reader should refer to appropriate bibliographies, a few of which are listed at the end of the volume.

I would like to express sincere thanks to Professor Richard Bellman for his suggestion to write this book. I am also indebted to Professor Akira Nakamura for his continual encouragement and useful discussions.

1

Introduction and Outline

1.1 Introduction

A history of the development of the mathematics of solitons begins in 1967 with a remarkable discovery by Gardner *et al.* [1] of an exact method for solving the initial value problem of the Korteweg–de Vries (KdV) equation. They reduced the *nonlinear* problem to the *linear* one, which was well known as the Sturm–Liouville eigenvalue problem characterized by the Schrödinger equation, and then discussed the properties of the exact solution describing the interaction of solitons.

Then this method, which we shall call the *inverse scattering method*, was extended to a more general form to be applicable to a wide class of nonlinear evolution equations such as the modified KdV equation, the nonlinear Schrödinger equation, and the Sine–Gordon equation

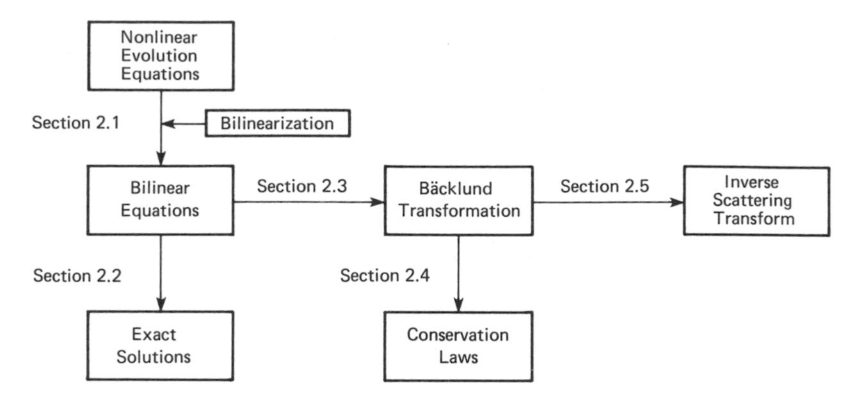

Fig. 1.1 Schematic illustration of the bilinear transformation method.

[2–4]. The inverse scattering method is now included in several textbooks. (See, for example, Bullough and Caudrey [5], Lamb [6], Ablowitz and Segur [7], and Calogero and Degasperis [8].)

In 1971 Hirota [9] developed an ingenious method for obtaining the exact multisoliton solution of the KdV equation and derived an explicit expression of the N-soliton solution. His method consisted of transforming the nonlinear evolution equation into the bilinear equation through the dependent variable transformation. The bilinear equation thus obtained can be solved by employing a perturbation method.[§]

The Bäcklund transformation [11, 12] is another method for finding multisoliton solutions of some class of nonlinear evolution equations. With this method, the multisoliton solutions can be constructed by purely algebraic procedures. Also, this method can be used to derive an infinite number of conservation laws. Hirota [13] introduced new bilinear operators together with their properties and then developed a unified method for constructing the Bäcklund transformation on the basis of the bilinear equation written in terms of new bilinear operators. At the same time Hirota clarified the relation between the Bäcklund transformation method and the inverse scattering method using his new formalism. The procedures mentioned above are depicted in Fig. 1.1.

[§] This method was shown to be applicable to a large class of nonlinear evolution equations [5, 9, 10].

1.2 Outline

In Chapter 2 the bilinear transformation method is applied to the KdV equation to illustrate this method. First, the KdV equation is bilinearized through a dependent variable transformation (Section 2.1). Then the methods for obtaining the N-soliton solution, the generalized soliton solution, which may be interpreted as a generalization of the N-soliton solution, and the periodic wave solution are presented (Section 2.2). Starting with the bilinearized KdV equation, the procedures to derive the Bäcklund transformation (Section 2.3), an infinite number of conservation laws (Section 2.4), and the inverse scattering transform (Section 2.5) are briefly discussed. The final section (Section 2.6) is devoted to the bilinear transformation method bibliography.

In Chapter 3 we discuss in detail the mathematical structure of the Benjamin–Ono (BO) equation from the viewpoint of three different methods: the bilinear transformation method, the theory of linear algebraic equation, and the pole expansion method. Physically, the BO equation describes a large class of internal waves in stratified fluids of great depth [14–17] and also governs the propagation of nonlinear Rossby waves in a rotating fluid [18]. Mathematically, the BO equation is a nonlinear *integrodifferential* equation with a dispersion term characterized by the Hilbert transform. Owing to this definite integral term, the solutions have many different properties from those of the well-known KdV type.

The N-soliton and N-periodic wave solutions are presented explicitly using the bilinear transformation method (Section 3.1.1). It is then demonstrated that the N-soliton solution can also be obtained by means of two different methods: the theory of linear algebraic equation (Section 3.1.2) and the pole expansion method (Section 3.1.3). The Bäcklund transformation is constructed in the bilinear form (Section 3.2.1) and is used to derive an infinite number of conservation laws (Section 3.2.2). A method for solving the initial value problem of the BO equation is then developed (Section 3.3.1) and the asymptotic solutions for large values of time are derived using a zero dispersion limit (Section 3.3.2). To apply this method, an initial condition evolving into pure N solitons is presented (Section 3.3.3). The stability of solitons is also discussed in relation to a small perturbation in the initial condition (Section 3.4). Finally, the initial value problem of the linearized BO equation is solved exactly, and asymptotic behaviors of solutions for large values of time are investigated (Section 3.5).

In Chapter 4 the general nature of the interaction of the BO solitons is investigated by employing the expression of the N-soliton solution (Section 4.1). The interaction of two solitons is then studied in detail (Section 4.2).

In Chapter 5 the BO-related equations are bilinearized using the bilinear transformation method and their solutions are presented. The equations treated are the Lax hierarchy of the BO equation [19–21], which we shall call the higher-order BO equations (Section 5.1), the higher-order KdV equations [22] (Section 5.2), the finite-depth fluid equation [23–28] and its higher-order equations [21, 29] (Section 5.3), and the higher-order modified KdV equations [30] (Section 5.4). The finite-depth fluid equation, which describes long waves in a stratified fluid of finite depth, is especially interesting since it reduces to the BO equation in the deep-water limit and to the KdV equation in the shallow-water limit and it therefore shares many of the properties of the BO and KdV equations. Finally, the Bäcklund transformations and the inverse scattering transforms of the higher-order KdV equations are constructed in the bilinear forms [31] (Section 5.5).

In Chapter 6 we treat other interesting topics related to the BO equation. The modified BO equation [32], which is generated from the Bäcklund transformation, is bilinearized, and the N-soliton and N-periodic wave solutions are presented (Section 6.1). The nonlinear Schrödinger equation, which derives from the nonlinear self-modulation problem of the BO equation [33, 34], is then discussed (Section 6.2). Finally, the effect of a small dissipation on the BO equation is considered. The system of equations that govern the time evolutions of the amplitudes and phases of the BO solitons is derived on the basis of the multiple time-scale expansion method (Section 6.3).

The Appendixes contain the properties of the bilinear operators together with their proofs (Appendix I[§]), the properties of the matrix that appears in the expression of the N-soliton solution of the BO equation (Appendix II), the properties of the Hilbert transform operator (Appendix III), and a proof of an exact solution of the integral equation that determines the distribution of the amplitudes of the BO solitons (Appendix IV).

[§] Appendix I will be very useful for the reader unfamiliar with the bilinear operators.

2

Introduction to the Bilinear Transformation Method

In this chapter the bilinear transformation method is explicitly illustrated by its application to the Korteweg–de Vries (KdV) equation, the prototype of the nonlinear evolution equation. The KdV equation was first derived by Korteweg and de Vries while developing a theory for shallow-water waves [35]; it later became clear that many physical systems could be described by the KdV equation [36–38]. The method of exact solution, the Bäcklund transformation, and the inverse scattering transform are based on the bilinearized KdV equation. The Bibliography (Section 2.6) discusses applications of the bilinear transformation method to other nonlinear equations.

2.1 Bilinearization

Let us consider the KdV equation in the form

$$u_t + 6uu_x + u_{xxx} = 0, \tag{2.1}$$

with the boundary condition $u \to 0$ as $|x| \to \infty$. Here $u = u(x, t)$ is a real function of both time t and space coordinate x, and subscripts denote partial differentiation.

What dependent variable transformation could be introduced to transform (2.1) into a more tractable form? A key is the steady-state solution of (2.1), that is,

$$u(x, t) = (p^2/2)\,\text{sech}^2(\eta/2), \tag{2.2}$$

where

$$\eta = px - p^3 t + \eta_0 \tag{2.3}$$

and p and η_0 are arbitrary constants. We can rewrite (2.2) in the form

$$u(x, t) = 2p^2(e^{\eta/2} + e^{-\eta/2})^{-2} = 2\,\partial^2 \ln(1 + e^{\eta})/\partial x^2. \tag{2.4}$$

The functional form of (2.4) suggests the following dependent variable transformation:

$$u(x, t) = 2\,\partial^2 \ln f(x, t)/\partial x^2. \tag{2.5}$$

Substituting (2.5) into (2.1) and integrating with respect to x, we obtain

$$f_{xt} f - f_x f_t + f_{xxxx} f - 4f_{xxx} f_x + 3(f_{xx})^2 = 0, \tag{2.6}$$

with the integration constant being set to zero. Equation (2.6) is the original version of the bilinearized KdV equation derived by Hirota [9]. It can be confirmed by a simple calculation that the function $f = 1 + e^{\eta}$ satisfies (2.6). We may regard f as a more fundamental quantity than u in the structure of the KdV equation.

We now introduce Hirota's bilinear operators defined by the following rule [13]:

$$D_t^n D_x^m a \cdot b = (\partial/\partial t - \partial/\partial t')^n (\partial/\partial x - \partial/\partial x')^m a(x, t) b(x', t')\Big|_{\substack{x'=x,\\ t'=t}} \tag{2.7}$$

where n and m are arbitrary nonnegative integers. Equation (2.6) can be compactly rewritten in terms of these bilinear operators as

$$D_x(D_t + D_x^3) f \cdot f = 0, \tag{2.8}$$

which is a convenient form to use in discussing exact solutions and the Bäcklund transformation of the KdV equation. The process of transforming the nonlinear evolution equation (2.1) into the form in (2.8) is called *bilinearization.*

If the boundary condition is given by $u \to u_0$ (= constant) as $|x| \to \infty$, we introduce the dependent variable transformation

$$u(x, t) = u_0 + 2\, \partial^2 \ln f(x, t)/\partial x^2 \tag{2.9}$$

in (2.1), yielding the bilinear equation

$$(D_t D_x + 6u_0 D_x^2 + D_x^4 + c) f \cdot f = 0, \tag{2.10}$$

where c is an integration constant. The bilinearized KdV equation, (2.8) or (2.10), is the starting point for the following sections.

Having demonstrated Hirota's method for bilinearizing the KdV equation, a typical nonlinear evolution equation, the question may arise as to what conditions allow for the bilinearization of a given nonlinear evolution equation. This question has not been answered in general, though many important nonlinear equations in physics and applied mathematics have already been bilinearized [10, 39]. The inverse problem, that of reducing a given bilinear equation to a nonlinear equation in the original variable u, is more tractable. For example, consider generalizations of (2.8) and (2.5):

$$D_x(D_t + D_x^5) f \cdot f = 0, \qquad u = 2\, \partial^2 \ln f/\partial x^2. \tag{2.11}$$

Using appendix formulas (App. I.3.4) and (App. I.5.3), Eqs. (2.11) are transformed as

$$u_t + 45u^2 u_x + 15u_x u_{xx} + 15uu_{xxx} + u_{xxxxx} = 0, \tag{2.12}$$

which is called the Sawada–Kotera equation [40]. We can similarly construct a wide variety of nonlinear evolution equations by reducing bilinear equations through dependent variable transformations.

The usefulness of the bilinear transformation method is related to the structure of nonlinear evolution equations; this method provides a simple, straightforward way of obtaining various types of exact solutions, which will be demonstrated in the next section.

2.2 Exact Solutions

2.2.1 Soliton Solution

We shall first develop a method for obtaining the N-soliton solution of the KdV equation on the basis of the bilinearized KdV equation (2.8) [11]. Let us expand f formally in powers of an arbitrary parameter ε as

$$f = \sum_{n=1}^{\infty} f_n \varepsilon^n, \qquad f_0 = 1. \tag{2.13}$$

Substituting (2.13) into (2.8) we obtain

$$\begin{aligned}\varepsilon F(D_t, D_x)(f_1 \cdot 1 + 1 \cdot f_1) &+ \varepsilon^2 F(D_t, D_x)(f_2 \cdot 1 + f_1 \cdot f_1 + 1 \cdot f_2) \\ &+ \varepsilon^3 F(D_t, D_x)(f_3 \cdot 1 + f_2 \cdot f_1 + f_1 \cdot f_2 + 1 \cdot f_3) + \cdots \\ &+ \varepsilon^n F(D_t, D_x)\left(\sum_{j=0}^{n} f_{n-j} \cdot f_n\right) + \cdots = 0,\end{aligned} \tag{2.14}$$

where we have set

$$F(D_t, D_x) = D_x(D_t + D_x^3) \tag{2.15}$$

for simplicity. Equating the coefficients ε^n $(n = 1, 2, \ldots)$ to zero on the left-hand side of (2.14) yields a hierarchy of equations for $f_j (j = 1, 2, \ldots)$.§ The first is

$$F(D_t, D_x)(f_1 \cdot 1 + 1 \cdot f_1) = 0, \tag{2.16}$$

which reduces to the linear equation

$$(\partial/\partial t + \partial^3/\partial x^3) f_1 = 0, \tag{2.17}$$

from (2.15) and properties of the bilinear operators (App. I.1.1), (App. I.1.2), and (App. I.1.4).

The simplest solution of (2.17) takes the form

$$f_1 = e^{\eta_1}, \qquad \eta_1 = p_1 x + \Omega_1 t + \eta_{01}, \tag{2.18}$$

§ Note that the equation derived from the ε^j term is linear with respect to f_j.

with p_1 and η_{01} being constants and $\Omega_1 = -p_1^3$. The ε^2 term of (2.14) and (2.18) gives

$$2F(D_t, D_x)f_2 \cdot 1 = -F(D_t, D_x)f_1 \cdot f_1 = -F(D_t, D_x)e^{\eta_1} \cdot e^{\eta_1} = 0, \tag{2.19}$$

using (App. I.2.1). Therefore, we may set $f_2 = 0$. Similarly, the equations corresponding to the ε^j $(j \geq 3)$ terms are satisfied by $f_j = 0$ $(j \geq 3)$. This implies that $1 + \varepsilon e^{\eta_1}$ is an exact solution of (2.8). If we set

$$\varepsilon = e^{\eta_0 - \eta_{01}}, \qquad p_1 = p \tag{2.20}$$

in (2.18), the result is equivalent to (2.4), the one-soliton solution of the KdV equation.

The exact solution of (2.17) may now be given as

$$f_1 = e^{\eta_1} + e^{\eta_2}, \tag{2.21}$$

with

$$\eta_j = p_j x + \Omega_j t + \eta_{0j}, \qquad \Omega_j = -p_j^3, \quad j = 1, 2, \tag{2.22}$$

where p_j and η_{0j} $(j = 1, 2)$ are constants. The equation for f_2 is given by

$$\begin{aligned} 2F(D_t, D_x)f_2 \cdot 1 &= -F(D_t, D_x)(e^{\eta_1} + e^{\eta_2}) \cdot (e^{\eta_1} + e^{\eta_2}) \\ &= -F(D_t, D_x)(e^{\eta_1} \cdot e^{\eta_2} + e^{\eta_2} \cdot e^{\eta_1}) \\ &= -2\frac{F(\Omega_1 - \Omega_2, p_1 - p_2)}{F(\Omega_1 + \Omega_2, p_1 + p_2)} F(D_t, D_x)e^{\eta_1 + \eta_2} \cdot 1, \end{aligned} \tag{2.23}$$

where use has been made of formula (App. I.2.2). From (2.23) f_2 is solved as

$$f_2 = -\frac{F(\Omega_1 - \Omega_2, p_1 - p_2)}{F(\Omega_1 + \Omega_2, p_1 + p_2)} e^{\eta_1 + \eta_2}. \tag{2.24}$$

Since $f_2 \neq 0$, we proceed to the ε^3 term in (2.14),

$$2F(D_t, D_x)f_3 \cdot 1 = -F(D_t, D_x)(f_2 \cdot f_1 + f_1 \cdot f_2). \tag{2.25}$$

Substituting (2.21) and (2.24) into (2.25) and using (App. I.2.2), we obtain

$$
\begin{aligned}
&2F(D_t, D_x) f_3 \cdot 1 \\
&= 2\frac{F(\Omega_1 - \Omega_2, P_1 - p_2)}{F(\Omega_1 + \Omega_2, p_1 + p_2)} F(D_t, D_x)\{e^{\eta_1+\eta_2} \cdot (e^{\eta_1} + e^{\eta_2}) \\
&\quad + (e^{\eta_1} + e^{\eta_2}) \cdot e^{\eta_1+\eta_2}\} \\
&= 2\frac{F(\Omega_1 - \Omega_2, p_1 - p_2)}{F(\Omega_1 + \Omega_2, p_1 + p_2)} F(D_t, D_x)(e^{\eta_1+\eta_2} \cdot e^{\eta_1} \\
&\quad + e^{\eta_1+\eta_2} \cdot e^{\eta_2} + e^{\eta_1} \cdot e^{\eta_1+\eta_2} + e^{\eta_2} \cdot e^{\eta_1+\eta_2}) \\
&= 2\frac{F(\Omega_1 - \Omega_2, p_1 - p_2)}{F(\Omega_1 + \Omega_2, p_1 + p_2)} \\
&\quad \times \left[\left\{\frac{F(\Omega_2, p_2)}{F(2\Omega_1 + \Omega_2, 2p_1 + p_2)} + \frac{F(-\Omega_2, -p_2)}{F(2\Omega_1 + \Omega_2, 2p_1 + p_2)}\right\}\right. \\
&\quad \times F(D_t, D_x) e^{2\eta_1+\eta_2} \cdot 1 \\
&\quad + \left\{\frac{F(\Omega_1, p_1)}{F(\Omega_1 + 2\Omega_2, p_1 + 2p_2)} + \frac{F(-\Omega_1, -p_1)}{F(\Omega_1 + 2\Omega_2, p_1 + 2p_2)}\right\} \\
&\quad \left. \times F(D_t, D_x) e^{\eta_1+2\eta_2} \cdot 1\right] = 0,
\end{aligned}
\tag{2.26}
$$

where we have used the relation

$$
F(\Omega_j, p_j) = F(-\Omega_j, -p_j) = p_j(\Omega_j + p_j^3) = 0, \qquad j = 1, 2 \tag{2.27}
$$

[by (2.22)], in arriving at the final value (2.26). Therefore, we can set $f_3 = 0$; it follows from the equations for f_j ($j \geq 4$) that $f_j = 0$ ($j \geq 4$). The solution thus obtained

$$
\begin{aligned}
f &= 1 + \varepsilon f_1 + \varepsilon^2 f_2 \\
&= 1 + \varepsilon(e^{\eta_1} + e^{\eta_2}) - \varepsilon^2 \frac{F(\Omega_1 - \Omega_2, p_1 - p_2)}{F(\Omega_1 + \Omega_2, p_1 + p_2)} e^{\eta_1+\eta_2}
\end{aligned}
\tag{2.28}
$$

corresponds to the two-soliton solution of the KdV equation.[§] It can be seen from (2.26) that (2.27) is crucial to the proof. Equation (2.27) is

[§] Note that ε can be set to one by replacing η_j with $\eta_j - \ln \varepsilon$ ($j = 1, 2$).

actually the dispersion relation of the linearized KdV equation

$$u_t + u_{xxx} = 0, \tag{2.29}$$

a significant observation.

We now proceed to the N-soliton solution. A special solution of (2.17), analogous to the one- and two-soliton solutions, takes the form

$$f_1 = \sum_{j=1}^{N} e^{\eta_j}, \tag{2.30}$$

where

$$\eta_j = p_j x + \Omega_j t + \eta_{0j}, \qquad j = 1, 2, \ldots, N \tag{2.31}$$

and

$$\Omega_j = -p_j^3, \qquad j = 1, 2, \ldots, N. \tag{2.32}$$

Setting $\varepsilon = 1$, without loss of generality, the N-soliton solution is given as the compact form derived by Hirota [5]

$$f = \sum_{\mu=0,1} \exp\left(\sum_{j=1}^{N} \mu_j \eta_j + \sum_{j>k}^{(N)} \mu_j \mu_k A_{jk} \right) \tag{2.33}$$

with

$$e^{A_{jk}} = -\frac{F(\Omega_j - \Omega_k, p_j - p_k)}{F(\Omega_j + \Omega_k, p_j - p_k)} = \frac{(p_j - p_k)^2}{(p_j + p_k)^2}, \tag{2.34}$$

where $\sum_{\mu=0,1}$ indicates the summation over all possible combinations of $\mu_1 = 0, 1, \mu_2 = 0, 1, \ldots, \mu_N = 0, 1$ and $\sum_{j>k}^{(N)}$ means the summation over all possible combinations of N elements under the condition $j > k$.[§] It can be easily confirmed that, for $N = 2$, Eq. (2.33) reduces to the two-soliton solution (2.28).

The somewhat tedious proof of (2.33) is by mathematical induction. We shall demonstrate the proof since this is essential to the bilinear transformation method. Substituting (2.33) into (2.8) and using (App. I.2.1) yields

$$\sum_{\mu=0,1} \sum_{\mu'=0,1} F\left(\sum_{j=1}^{N} (\mu_j - \mu_j')\Omega_j, \sum_{j=1}^{N} (\mu_j - \mu_j')p_j \right)$$

$$\times \exp\left[\sum_{j=1}^{N} (\mu_j + \mu_j')\eta_j + \sum_{j>k}^{(N)} (\mu_j \mu_k + \mu_j' \mu_k')A_{jk} \right] = 0, \tag{2.35}$$

where F is given by (2.15). Let the coefficient of the factor

$$\exp\left(\sum_{j=1}^{n} \eta_j + 2 \sum_{j=n+1}^{m} \eta_j \right)$$

[§] These notations will be used throughout this book.

on the left-hand side of (2.35) be G. It follows that

$$G = \sum_{\mu=0,1} \sum_{\mu'=0,1} \text{cond}(\mu, \mu') F\left(\sum_{j=1}^{N} (\mu_j - \mu'_j)\Omega_j, \sum_{j=1}^{N} (\mu_j - \mu'_j) p_j \right) \times \exp\left[\sum_{j>k}^{(N)} (\mu_j \mu_k + \mu'_j \mu'_k) A_{jk} \right], \tag{2.36}$$

where the notation cond(μ, μ') implies that summations over μ and μ' are performed under the following conditions:

$$\begin{aligned} \mu_j + \mu'_j &= 1 \quad \text{for} \quad j = 1, 2, \ldots, n, \\ \mu_j = \mu'_j &= 1 \quad \text{for} \quad j = n+1, n+2, \ldots, m, \\ \mu_j = \mu'_j &= 0 \quad \text{for} \quad j = m+1, m+2, \ldots, N. \end{aligned} \tag{2.37}$$

Defining the variable

$$\sigma_j = \mu_j - \mu'_j, \tag{2.38}$$

we have

$$\begin{aligned} &\sum_{j>k}^{(N)} (\mu_j \mu_k + \mu'_j \mu'_k) A_{jk} \\ &= \frac{1}{2} \sum_{j,k=1}^{N} (\mu_j \mu_k + \mu'_j \mu'_k) A_{jk} - \sum_{j=1}^{N} (\mu_j^2 + \mu_j'^2) A_{jj} \\ &= \frac{1}{2} \left[\sum_{j=1}^{n} \sum_{k=1}^{n} + \sum_{j=1}^{n} \sum_{k=n+1}^{m} + \sum_{j=1}^{n} \sum_{k=m+1}^{N} + \sum_{j=n+1}^{m} \sum_{k=1}^{n} \right. \\ &\quad + \sum_{j=n+1}^{m} \sum_{k=n+1}^{m} + \sum_{j=n+1}^{m} \sum_{k=m+1}^{N} \\ &\quad \left. + \sum_{j=m+1}^{N} \sum_{k=1}^{n} + \sum_{j=m+1}^{N} \sum_{k=n+1}^{m} + \sum_{j=n+1}^{N} \sum_{k=m+1}^{N} \right] (\mu_j \mu_k + \mu'_j \mu'_k) A_{jk} \\ &= \frac{1}{2} \sum_{j,k=1}^{n} \left[\frac{1}{4}(1+\sigma_j)(1+\sigma_k) + \frac{1}{4}(1-\sigma_j)(1-\sigma_k) \right] A_{jk} \\ &\quad + \sum_{j=1}^{n} \sum_{k=n+1}^{m} A_{jk} + \sum_{j,k=n+1}^{m} A_{jk} \\ &= \sum_{j>k}^{(n)} \frac{1}{2}(1 + \sigma_j \sigma_k) A_{jk} + \sum_{j=1}^{n} \sum_{k=n+1}^{m} A_{jk} + \sum_{j,k=n+1}^{m} A_{jk}, \end{aligned} \tag{2.39}$$

using (2.37) and setting $A_{jj} = 0$, a consequence of (2.34).

Since σ_j and σ_k take the values $+1$ or -1 for $1 \le j, k \le n$ [by (2.37) and (2.38)], we obtain, from the relations $F(\Omega, p) = F(-\Omega, -p)$ and (2.34),

$$\exp[\tfrac{1}{2}(1 + \sigma_j\sigma_k)A_{jk}] = -\frac{F(\sigma_j\Omega_j - \sigma_k\Omega_k, \sigma_j p_j - \sigma_k p_k)}{F(\Omega_j + \Omega_k, p_j + p_k)}\sigma_j\sigma_k. \quad (2.40)$$

Substituting (2.38), (2.39), and (2.40) into (2.36) yields

$$G = c\sum_{\sigma=\pm 1} F\left(\sum_{j=1}^{n}\sigma_j\Omega_j, \sum_{j=1}^{n}\sigma_j p_j\right)\prod_{j>k}^{(n)} F(\sigma_j\Omega_j - \sigma_k\Omega_k, \sigma_j p_j - \sigma_k p_k)\sigma_j\sigma_k, \quad (2.41)$$

where c is a constant that is independent of the summation indices $\sigma_1, \sigma_2, \ldots, \sigma_N$. If we can verify the identity

$$\sum_{\sigma=\pm 1} F\left(\sum_{j=1}^{n}\sigma_j\Omega_j, \sum_{j=1}^{n}\sigma_j p_j\right)\prod_{j>k}^{(n)} F(\sigma_j\Omega_j - \sigma_k\Omega_k, \sigma_j p_j - \sigma_k p_k)\sigma_j\sigma_k = 0 \quad (2.42)$$

for $n = 1, 2, \ldots, N$, then (2.33) is an exact solution of the KdV equation. Using (2.15) and (2.32), (2.42) becomes

$$\begin{aligned} G^{(n)}&(p_1, p_2, \ldots, p_n) \\ &\equiv \sum_{\sigma=\pm 1}\left(\sum_{j=1}^{n}\sigma_j p_j\right)\left\{-\sum_{j=1}^{n}(\sigma_j p_j)^3 + \left(\sum_{j=1}^{n}\sigma_j p_j\right)^3\right\}\prod_{j>k}^{(n)}(\sigma_j p_j - \sigma_k p_k)^2 \\ &\equiv 0. \end{aligned} \quad (2.43)$$

We shall now prove (2.43) by mathematical induction. For $n = 1, 2$, (2.43) obviously holds. If we assume (2.43) to be true up to $n - 1$, then

$$G^{(n)}(p_1, p_2, \ldots, p_n)|_{p_1=0} = 2\prod_{j=2}^{n} p_j^2 G^{(n-1)}(p_2, p_3, \ldots, p_n) = 0, \quad (2.44)$$

by induction. Furthermore,

$$\begin{aligned} G^{(n)}(p_1, p_2, \ldots, p_n)|_{p_1=\pm p_2} &= 8p_1^2\prod_{j=3}^{n}(p_1^2 - p_j^2)^2 G^{(n-2)}(p_3, p_4, \ldots, p_n) \\ &= 0, \end{aligned} \quad (2.45)$$

again by induction. Owing to the σ summation, $G^{(n)}$ is an even function of $p_1, p_2, \ldots, p_n$ and invariant under the replacement of p_j and p_k for arbitrary j and k. These properties, together with (2.44) and (2.45), lead to the factorization of $G^{(n)}$ as

$$G^{(n)}(p_1, p_2, \ldots, p_n) = \prod_{j=1}^{n} p_j^2 \prod_{j>k}^{(n)} (p_j^2 - p_k^2)^2 \tilde{G}(p_1, p_2, \ldots, p_n), \quad (2.46)$$

where $\tilde{G}$ is a polynomial of $p_1, p_2, \ldots, p_n$. Expression (2.46) shows that the degree of $G^{(n)}$ with respect to $p_1, p_2, \ldots, p_n$ is at least $2n(n-1) + 2n = 2n^2$. We see from (2.43) that the degree of $G^{(n)}$ is at most $n(n-1) + 4$. This is impossible for $n \geq 2$. Hence $G^{(n)}$ must be zero identically, completing the proof.

We now consider Hirota's theorem [5]. Consider the following form of the bilinear equation:

$$F(D_t, D_x) f \cdot f = 0, \quad (2.47)$$

where F is a polynomial or exponential function of D_t and D_x and satisfies the conditions

$$F(D_t, D_x) = F(-D_t, -D_x) \quad (2.48)$$

$$F(0, 0) = 0. \quad (2.49)$$

Note that the bilinearized KdV equation (2.8) is a special case of (2.47) using (2.48) and (2.49). Hirota's theorem states that the expression

$$f = \sum_{\mu=0,1} \exp\left(\sum_{j=1}^{N} \mu_j \eta_j + \sum_{j>k}^{(N)} \mu_j \mu_k A_{jk} \right), \quad (2.50)$$

with

$$\eta_j = p_j x + \Omega_j t + \eta_{0j}, \qquad j = 1, 2, \ldots, N, \quad (2.51)$$

$$F(\Omega_j, p_j) = 0, \qquad j = 1, 2, \ldots, N, \quad (2.52)$$

$$e^{A_{jk}} = -\frac{F(\Omega_j - \Omega_k, p_j - p_k)}{F(\Omega_j + \Omega_k, p_j + p_k)}, \qquad j, k = 1, 2, \ldots, N, \quad (2.53)$$

gives the N-soliton solution of (2.47) provided identity (2.42) holds for $n = 1, 2, \ldots, N$. [Here again we stress the importance of the dispersion relation (2.52).] The proof of Hirota's theorem is of the same form as that presented for the KdV equation.

For $N = 2$ the left-hand side of (2.42) becomes

$$\begin{aligned}
&\sum_{\sigma_1=\pm 1}\sum_{\sigma_2=\pm 1} F(\sigma_1\Omega_1 + \sigma_2\Omega_2, \sigma_1 p_1 + \sigma_2 p_2) \\
&\qquad\qquad \times F(\sigma_2\Omega_2 - \sigma_1\Omega_1, \sigma_2 p_2 - \sigma_1 p_1)\sigma_2\sigma_1 \\
&\quad = F(\Omega_1 + \Omega_2, p_1 + p_2)F(\Omega_2 - \Omega_1, p_2 - p_1) \\
&\qquad - F(\Omega_1 - \Omega_2, p_1 - p_2)F(-\Omega_2 - \Omega_1, -p_1 - p_2) \\
&\qquad - F(-\Omega_1 + \Omega_2, -p_1 + p_2)F(\Omega_2 + \Omega_1, p_2 + p_1) \\
&\qquad + F(-\Omega_1 - \Omega_2, -p_1 - p_2)F(-\Omega_2 + \Omega_1, -p_2 + p_1) \\
&\quad = 0,
\end{aligned} \tag{2.54}$$

using condition (2.48). It is concluded from this fact that *the two-soliton solution always exists without further conditions on F.*

Let us again consider the KdV equation. The N-soliton solution of the KdV equation (2.33) has interesting properties, one of which is its asymptotic behavior for large values of time [41–43]. To see this we order the parameters p_j ($j = 1, 2, \ldots, N$) as

$$0 < p_1 < p_2 < \cdots < p_N \tag{2.55}$$

and transform the reference frame, which moves with the velocity p_n^2 ($n = 1, 2, \ldots,$ or N), giving

$$\xi = x - p_n^2 t. \tag{2.56}$$

Noting that

$$\begin{aligned}
\eta_j &= p_j(x - p_j^2 t) + \eta_{0j} \\
&= p_j(p_n^2 - p_j^2)t + p_j\xi + \eta_{0j}, \qquad j = 1, 2, \ldots, N,
\end{aligned} \tag{2.57}$$

and using (2.55), in the limit of $t \to +\infty$

$$\begin{aligned}
f &= \exp\left(\sum_{j=1}^{n}\eta_j + \frac{1}{2}\sum_{j=1}^{n}\sum_{k=1}^{n}A_{jk}\right) \\
&\quad + \exp\left(\sum_{j=1}^{n-1}\eta_j + \frac{1}{2}\sum_{j=1}^{n-1}\sum_{k=1}^{n-1}A_{jk}\right) + O\left[\exp\left(\sum_{j=1}^{n-2}\eta_j\right)\right] \\
&\simeq \exp\left(\sum_{j=1}^{n}\eta_j + \frac{1}{2}\sum_{j=1}^{n}\sum_{k=1}^{n}A_{jk}\right)\left[1 + \exp\left(-\eta_n - \sum_{j=1}^{n-1}A_{nj}\right)\right].
\end{aligned} \tag{2.58}$$

Therefore,

$$u(x, t) = 2\,\partial^2 \ln f/\partial x^2$$
$$= (p_n^2/2)\,\mathrm{sech}^2[(p_n/2)(x - p_n^2 t + \xi_n^+) + \eta_{0n}], \qquad t \to +\infty, \tag{2.59}$$

with

$$\xi_n^+ = \frac{1}{p_n}\sum_{j=1}^{n-1} A_{nj}. \tag{2.60}$$

Similarly, in the limit of $t \to -\infty$,

$$u(x, t) = (p_n^2/2)\,\mathrm{sech}^2[(p_n/2)(x - p_n^2 t + \xi_n^-) + \eta_{0n}], \qquad t \to -\infty, \tag{2.61}$$

with

$$\xi_n^- = \frac{1}{p_n}\sum_{j=n+1}^{N} A_{nj}. \tag{2.62}$$

In view of (2.34)

$$A_{nj} = 2\ln\left|\frac{p_n - p_j}{p_n + p_j}\right|, \qquad n \neq j. \tag{2.63}$$

The functional form of (2.59) [or (2.61)] is exactly the same as that for the one-soliton solution (2.2) except for the phase shift ξ_n^+ (or ξ_n^-). In other words, the N-soliton solution evolves asymptotically as $t \to \pm\infty$ into localized solitons moving with constant velocities $p_1^2, p_2^2, \ldots, p_N^2$ in the original reference frame. As $t \to +\infty$, the trajectory of the nth soliton is shifted by the quantity

$$\xi_n^+ - \xi_n^- = \frac{2}{p_n}\left[\sum_{j=1}^{n-1}\ln\left(\frac{p_n - p_j}{p_n + p_j}\right) - \sum_{j=n+1}^{N}\ln\left(\frac{p_j - p_n}{p_j + p_n}\right)\right], \tag{2.64}$$

relative to that for $t \to -\infty$. It can be seen from this expression that the interaction between solitons occurs only in a pairwise way. It also follows from (2.64) that

$$\sum_{n=1}^{N} p_n \xi_n^+ = \sum_{n=1}^{N} p_n \xi_n^-, \tag{2.65}$$

implying the conservation of the total phase shift.

Finally, we note that expression (2.33) of the N-soliton solution can be rewritten in determinant form as

$$f = \det M, \tag{2.66}$$

where M is an $N \times N$ matrix whose elements are given by

$$M_{jk} = \begin{cases} 1 + e^{\eta_j} & \text{for} \quad j = k, \qquad (2.67) \\ \dfrac{2(p_j p_k)^{1/2}}{p_j + p_k} e^{(\eta_j + \eta_k)/2} & \text{for} \quad j \neq k. \qquad (2.68) \end{cases}$$

Further details of the properties of the N-soliton solution of the KdV equation can be found in the references [41–43].

2.2.2 *Generalized Soliton Solution*

The N-soliton solution that we have discussed is a special solution of the nonlinear evolution equation. We shall now develop a method for constructing a more general solution (which we shall call a generalized soliton solution) that describes interactions between solitons and ripples. This method of solution, primarily developed by Rosales [44] and Oishi [45–47], is an extension of the bilinear transformation method, and we shall illustrate it in the case of the KdV equation, demonstrating that the generalized soliton solution solves the initial value problem of the KdV equation.

As shown in (2.14), if functions f_n ($n = 0, 1, 2, \ldots$) defined in (2.13) satisfy the system of equations

$$\sum_{s=0}^{n} F(D_t, D_x) f_{n-s} \cdot f_s = 0, \qquad n = 0, 1, 2, \ldots, \tag{2.69}$$

with

$$F(D_t, D_x) = D_x(D_t + D_x^3), \tag{2.70}$$

then the f given by (2.13) is an exact solution of the KdV equation.[§]

For $n = 1$, (2.69) becomes

$$F(D_t, D_x) f_1 \cdot 1 = 0, \tag{2.71}$$

where we have used the property

$$F(D_t, D_x) f \cdot g = F(D_t, D_x) g \cdot f. \tag{2.72}$$

Obviously (2.71) is identified with the linear equation

$$f_{1,t} + f_{1,xxx} = 0 \tag{2.73}$$

after integration with respect to x. The solution of (2.73) is readily obtained in an integral form as

$$f_1(x, t) = \int_\Gamma \exp(px - p^3 t)\, d\tau(p). \tag{2.74}$$

[§] Note that (2.69) is a linear equation for f_n.

Here $\int_\Gamma d\tau(p)$ denotes the contour integral along the contour Γ which lies in the left half of the complex p plane (that is, $\text{Re}\, p < 0$) and which goes from $p = -i\infty$ to $p = i\infty$. If the measure $d\tau(p)$ introduced in (2.74) is chosen as

$$\int_\Gamma \exp(px - p^3 t)\, d\tau(p) = \int_{-\infty}^{\infty} c(p_r) \exp(ip_r x - ip_r^3 t)\, dp_r + \sum_{j=1}^{N} c_j \exp(p_{0j} x - p_{0j}^3 t), \tag{2.75}$$

where $c(p_r)$ is a real function of p_r and p_{0j} $(j = 1, 2, \ldots, N)$ are real constants, then the first term on the right-hand side of (2.75) represents ripples (or dispersive waves) and the second term represents solitons. Therefore, it may be called the generalized soliton solution.

For $n = 2$, using (2.69) and (2.74),

$$\begin{aligned} F(D_t, D_x) f_2 \cdot 1 &= -\frac{1}{2} F(D_t, D_x) f_1 \cdot f_1 \\ &= -\frac{1}{2} \int_\Gamma \int_\Gamma F(D_t, D_x) \exp(p_1 x - p_1^3 t) \\ &\quad \cdot \exp(p_2 x - p_2^3 t)\, d\tau(p_1)\, d\tau(p_2) \\ &= -\frac{1}{2} \int_\Gamma \int_\Gamma F(-p_1^3 + p_2^3, p_1 - p_2) \\ &\quad \times \exp[(p_1 + p_2)x - (p_1^3 + p_2^3)t]\, d\tau(p_1)\, d\tau(p_2), \end{aligned} \tag{2.76}$$

where use has been made of formula (App. I.2.1). Noting the relations

$$F(D_t, D_x) f_2 \cdot 1 = f_{2,tx} + f_{2,xxxx}, \tag{2.77}$$

$$\begin{aligned} F(-p_1^3 + p_2^3, p_1 - p_2) &= (p_1 - p_2)[-p_1^3 + p_2^3 + (p_1 - p_2)^3] \\ &= -3p_1 p_2 (p_1 - p_2)^2, \end{aligned} \tag{2.78}$$

$$\begin{aligned} &\left(\frac{\partial^2}{\partial t\, \partial x} + \frac{\partial^4}{\partial x^4}\right) \exp[(p_1 + p_2)x - (p_1^3 + p_2^3)t] \\ &\quad = 3p_1 p_2 (p_1 + p_2)^2 \exp[(p_1 + p_2)x - (p_1^3 + p_2^3)t], \end{aligned} \tag{2.79}$$

(2.76) is solved as

$$f_2 = \frac{1}{2} \int_\Gamma \int_\Gamma \left(\frac{p_1 - p_2}{p_1 + p_2}\right)^2 \exp\left[\sum_{j=1}^{2} (p_j x - p_j^3 t)\right] \prod_{j=1}^{2} d\tau(p_j). \tag{2.80}$$

In the same way, for general n, f_n is determined as [46]

$$f_n = \frac{1}{n!}\int_\Gamma\int_\Gamma\cdots\int_\Gamma \prod_{1\le j<k\le n}\left[-\frac{F(\Omega_j-\Omega_k, p_j-p_k)}{F(\Omega_j+\Omega_k, p_j+p_k)}\right]$$
$$\times \exp\left[\sum_{j=1}^{n}(p_jx-p_j^3t)\right]\prod_{j=1}^{n}d\tau(p_j),$$
$$= \frac{1}{n!}\int_\Gamma\int_\Gamma\cdots\int_\Gamma \prod_{1\le j<k\le n}\left(\frac{p_j-p_k}{p_j+p_k}\right)^2$$
$$\times \exp\left[\sum_{j=1}^{n}(p_jx-p_j^3t)\right]\prod_{j=1}^{n}d\tau(p_j), \qquad n\ge 1, \tag{2.81}$$

where $\Omega_j = -p_j^3$ $(j = 1, 2, \ldots, n)$. We now show that (2.81) satisfies (2.69). Substituing (2.81) into (2.69) yields

$$A \equiv \sum_{s=0}^{n} F(D_t, D_x) f_s\cdot f_{n-s}$$
$$= \sum_{s=0}^{n}\frac{1}{s!(n-s)!}\int_\Gamma\int_\Gamma\cdots\int_\Gamma \prod_{1\le j<k\le s}\left[-\frac{F(\Omega_j-\Omega_k, p_j-p_k)}{F(\Omega_j+\Omega_k, p_j+p_k)}\right]$$
$$\times \prod_{s+1\le j<k\le n}\left[-\frac{F(\Omega_j-\Omega_k, p_j-p_k)}{F(\Omega_j+\Omega_k, p_j+p_k)}\right]$$
$$\times F(D_t, D_x)\exp\left[\sum_{j=1}^{s}(p_jx+\Omega_jt)\right]$$
$$\cdot\exp\left[\sum_{j=s+1}^{n}(p_jx+\Omega_jt)\right]\prod_{j=1}^{n}d\tau(p_j)$$
$$= \frac{1}{n!}\sum_{s=0}^{n} {}_nC_s\int_\Gamma\int_\Gamma\cdots\int_\Gamma F\left(\sum_{j=1}^{s}\Omega_j - \sum_{j=s+1}^{n}\Omega_j, \sum_{j=1}^{s}p_j - \sum_{j=s+1}^{n}p_j\right)$$
$$\times \prod_{1\le j<k\le s}\left[-\frac{F(\Omega_j-\Omega_k, p_j-p_k)}{F(\Omega_j+\Omega_k, p_j+p_k)}\right]$$
$$\times \prod_{s+1\le j<k\le n}\left[-\frac{F(\Omega_j-\Omega_k, p_j-p_k)}{F(\Omega_j+\Omega_k, p_j+p_k)}\right]$$
$$\times \exp\left[\sum_{j=1}^{n}(p_jx+\Omega_jt)\right]\prod_{j=1}d\tau(p_j), \tag{2.82}$$

where we have used formula (App. I.2.1) and introduced the notation

$$ {}_nC_s = n!/(n-s)!s!. \tag{2.83}$$

Equation (2.82) can be transformed into the form

$$\begin{aligned} A = {} & \frac{1}{n!} \sum_{\sigma=\pm 1} \sum_{s=0}^{n} {}_nC_s \\ & \times \frac{(1+\sigma_1)(1+\sigma_2)\cdots(1+\sigma_s)(1-\sigma_{s+1})(1-\sigma_{s+2})\cdots(1-\sigma_n)}{2^n} \\ & \times \int_\Gamma \int_\Gamma \cdots \int_\Gamma F\left(\sum_{j=1}^{n} \sigma_j \Omega_j,\ \sum_{j=1}^{n} \sigma_j p_j\right) \\ & \times \prod_{1\le j<k\le n} \left[-\frac{F(\sigma_j\Omega_j - \sigma_k\Omega_k,\ \sigma_j p_j - \sigma_k p_k)}{F(\Omega_j+\Omega_k,\ p_j+p_k)} \sigma_j \sigma_k \right] \\ & \times \exp\left[\sum_{j=1}^{n} (p_j x + \Omega_j t)\right] \prod_{j=1}^{n} d\tau(p_j), \end{aligned} \tag{2.84}$$

where $\sum_{\sigma=\pm 1}$ means the summation over all possible combinations of $\sigma_1 = \pm 1, \sigma_2 = \pm 1, \ldots, \sigma_n = \pm 1$. The equivalence of (2.82) and (2.84) can be confirmed by noting that there is a contribution to the summation only when $\sigma_1 = \sigma_2 = \cdots = \sigma_s = 1$ and $\sigma_{s+1} = \sigma_{s+2} = \cdots = \sigma_n = -1$ and $F(D_t, D_x) = F(-D_t, -D_x)$. Since the integrand in (2.84), which we set as I for simplicity, is unchanged by replacing p_j with p_{m_j} and σ_j with σ_{m_j} ($j = 1, 2, \ldots, n$), we obtain

$$\begin{aligned} A = {} & \frac{1}{n!} \sum_{\sigma=\pm 1} \sum_{s=0}^{n} \left[\sum_{{}_nC_s} 2^{-n}(1+\sigma_{m_1})\cdots(1+\sigma_{m_s}) \right. \\ & \left. \times (1-\sigma_{m_{s+1}})\cdots(1-\sigma_{m_n}) \right] \\ & \times \int_\Gamma \int_\Gamma \cdots \int_\Gamma I \prod_{j=1}^{n} d\tau(p_j), \end{aligned} \tag{2.85}$$

where $\sum_{{}_nC_s}$ denotes the summation over all possible combinations of s elements taken from n. Using the identity

$$\begin{aligned}\sum_{s=0}^{n} \sum_{{}_nC_s} & 2^{-n}(1+\sigma_{m_1})\cdots(1+\sigma_{m_s})(1-\sigma_{m_{s+1}})\cdots(1-\sigma_{m_n}) \\ &= 2^{-n} \sum_{\varepsilon_1,\ldots,\varepsilon_n=\pm 1} (1+\varepsilon_1\sigma_1)(1+\varepsilon_2\sigma_2)\cdots(1+\varepsilon_n\sigma_n) \\ &= 2^{-n} \sum_{\varepsilon_2,\ldots,\varepsilon_n=\pm 1} [(1+\sigma_1)+(1-\sigma_1)](1+\varepsilon_2\sigma_2)\cdots(1+\varepsilon_n\sigma_n) \\ &= 2^{-(n-1)} \sum_{\varepsilon_2,\ldots,\varepsilon_n=\pm 1} (1+\varepsilon_2\sigma_2)\cdots(1+\varepsilon_n\sigma_n) = \cdots = 1, \end{aligned} \tag{2.86}$$

(2.85) becomes

$$\begin{aligned} A = \frac{1}{n!}\int_\Gamma\int_\Gamma\cdots\int_\Gamma \sum_{\sigma=\pm 1} & F\left(\sum_{j=1}^{n}\sigma_j\Omega_j, \sum_{j=1}^{n}\sigma_j p_j\right) \\ \times \prod_{1\le j<k\le n} & \left[-\frac{F(\sigma_j\Omega_j-\sigma_k\Omega_k, \sigma_j p_j-\sigma_k p_k)}{F(\Omega_j+\Omega_k, p_j+p_k)}\sigma_j\sigma_k\right] \\ \times \exp & \left[\sum_{j=1}^{n}(p_j x+\Omega_j t)\right]\prod_{j=1}^{n} d\tau(p_j), \end{aligned} \tag{2.87}$$

which vanishes due to identity (2.42). Thus, we have shown that f_n, given by (2.81), satisfies Eq. (2.69).

Although we have been concerned with the KdV equation, this method can be applied to a wide class of nonlinear evolution equations. It may be summarized as follows: If the bilinear equation

$$F(D_t, D_x)f\cdot f = 0, \tag{2.88}$$

with subsidiary conditions

$$F(D_t, D_x)f\cdot g = F(D_t, D_x)g\cdot f \tag{2.89}$$

$$F(D_t, D_x)1\cdot 1 = 0 \tag{2.90}$$

has an N-soliton solution or, in other words, operator F satisfies identity (2.42), then it also contains the generalized soliton solution given by (2.13) and (2.81), where the dispersion relation in (2.81) must be replaced by

$$F(\Omega_j, p_j) = 0, \qquad j = 1, 2, \ldots, n, \tag{2.91}$$

instead of $\Omega_j = -p_j^3$. The pure N-soliton solution is obtained from (2.81) by defining the measure $d\tau(p_j)$ as

$$d\tau(p_j) = \int_{-\infty}^{\infty} \sum_{s=1}^{N} c_s \delta(p_j - p_{0s})\, dp_j, \tag{2.92}$$

where c_s and p_{0s} ($s = 1, 2, \ldots, N$) are real constants with $p_{0j} \neq p_{0k}$ for $j \neq k$ ($1 \leq j, k \leq N$) and δ is Dirac's delta function. Substitution of (2.92) into (2.81) yields

$$\begin{aligned} f &= \sum_{n=0}^{\infty} f_n \varepsilon^n \\ &= \sum_{n=0}^{N} \frac{\varepsilon^n}{n!} \sum_{s_1, \ldots, s_n = 1}^{N} c_{s_1} \cdots c_{s_n} \prod_{1 \leq j < k \leq n} \left(\frac{p_{0s_j} - p_{0s_k}}{p_{0s_j} + p_{0s_k}} \right)^2 \\ &\quad \times \exp\left[\sum_{j=1}^{n} (p_{0s_j} x - p_{0s_j}^3 t) \right] \\ &= \sum_{n=0}^{N} \sum_{{}_N C_n} \prod_{1 \leq j < k \leq n} \left(\frac{p_{0s_j} - p_{0s_k}}{p_{0s_j} + p_{0s_k}} \right)^2 \exp\left(\sum_{j=1}^{n} \eta_{s_j} \right), \end{aligned} \tag{2.93}$$

where

$$\eta_{s_j} = p_{0s_j} x - p_{0s_j}^3 t + \ln \varepsilon c_{s_j}, \qquad j = 1, 2, \ldots, n. \tag{2.94}$$

Note that f_n with $n \geq N + 1$ vanishes identically owing to the factor

$$\prod_{1 \leq j < k \leq n} \left(\frac{p_{0s_j} - p_{0s_k}}{p_{0s_j} + p_{0s_k}} \right)^2, \tag{2.95}$$

so that the infinite sum reduces to the finite one. It can easily be shown that expression (2.93) coincides with the N-soliton solution (2.33) of the KdV equation.

Finally, we shall briefly discuss the initial value problem of the KdV equation. We may rewrite the generalized soliton solution of the KdV equation (2.81) in the form

$$f = 1 + \sum_{n=1}^{\infty} \frac{\varepsilon^n}{n!} \int_\Gamma \int_\Gamma \cdots \int_\Gamma \det(\Delta_n) \exp\left[\sum_{j=1}^{n} (p_j x - p_j^3 t) \right] \prod_{j=1}^{n} d\tau(p_j), \tag{2.96}$$

where Δ_n is an $n \times n$ matrix whose elements are given by

$$(\Delta_n)_{jk} = 2p_j/(p_j + p_k), \qquad j, k = 1, 2, \ldots, n. \tag{2.97}$$

The equivalence of (2.81) and (2.96) may be seen by noting the identity of the Gram determinant

$$\det \Delta_n = \prod_{1 \le j < k \le n} \left(\frac{p_j - p_k}{p_j + p_k} \right)^2. \tag{2.98}$$

Equation (2.96) may be rewritten as

$$f = 1 + \sum_{n=1}^{\infty} \frac{\varepsilon^n}{n!} \int_x^{\infty} \int_x^{\infty} \cdots \int_x^{\infty} \det(\Phi_n) \prod_{j=1}^{n} ds_j, \tag{2.99}$$

where Φ_n is an $n \times n$ matrix with the (j, k) element given by $F(s_j + s_k; t)$. Here

$$F(s; t) = -\int_{\Gamma} \left[\exp\left(\frac{ps}{2} - p^3 t \right) \right] p \, d\tau(p). \tag{2.100}$$

We now introduce function $D(x, z; t)$, which corresponds to the Fredholm first minor by the relation

$$D(x, z; t) = -F(x + z; t) - \sum_{n=1}^{\infty} \frac{\varepsilon^n}{n!} \int_x^{\infty} \int_x^{\infty} \cdots \int_x^{\infty} \det(\Omega_n) \prod_{j=1}^{n} ds_j, \tag{2.101}$$

where Ω_n is an $(n + 1) \times (n + 1)$ matrix given by

$$\Omega_n = \begin{bmatrix} F(x + z; t) & F(x + s_1; t) & F(x + s_2; t) & \cdots & F(x + s_n; t) \\ F(s_1 + z; t) & F(s_1 + s_1; t) & F(s_1 + s_2; t) & \cdots & F(s_1 + s_n; t) \\ \vdots & \vdots & \vdots & \ddots & \vdots \\ F(s_n + z; t) & F(s_n + s_1; t) & F(s_n + s_2; t) & \cdots & F(s_n + s_n; t) \end{bmatrix}. \tag{2.102}$$

Differentiating (2.99) with respect to x and comparing with (2.101), yields the useful expression

$$D(x, x; t) = f_x(x, t). \tag{2.103}$$

To obtain the relation between D and f we expand $\det(\Omega_n)$ with respect to the first column. It follows from (2.99) and (2.101) that

$$D(x, z; t) = -F(x + z; t) f(x, t) - \varepsilon \int_x^{\infty} D(x, s; t) F(s + z; t) \, ds. \tag{2.104}$$

If we define the function $K(x, z; t)$ as

$$K(x, z; t) = D(x, z; t)/f(x, t), \tag{2.105}$$

where $f(x, t) \neq 0$ is assumed, then $K(x, z; t)$ satisfies the Gel'fand–Levitan–Marchenko (GLM) integral equation

$$K(x, z; t) + F(x + z; t) + \varepsilon \int_x^{\infty} K(x, s; t)F(s + z; t)\, ds = 0. \tag{2.106}$$

Inversely, if $K(x, z; t)$ is a solution of (2.106), then $K(x, z; t)$ is represented as

$$K(x, x; t) = \frac{D(x, x; t)}{f(x, t)} = [\ln f(x, t)]_x. \tag{2.107}$$

Relation (2.107), combined with (2.5), yields the useful expression

$$u(x, t) = 2[K(x, x; t)]_x. \tag{2.108}$$

Function $K(x, z; t)$ is satisfied by the linear equation [47]

$$K_{xx}(x, z; t) - K_{zz}(x, z; t) + u(x, t)K(x, z; t) = 0 \tag{2.109}$$

by applying the uniqueness theorem to the solution of the GLM equation.

The procedure for solving the initial value problem of the KdV equation is now established and can be explained as follows: Given initial data $u(x, 0)$, first solve (2.109) under the boundary condition $K(x, z; 0) \to 0$ as $x + z \to \infty$, then introduce this $K(x, z; 0)$ into (2.106) to obtain $F(x; 0)$. Next, determine the measure $d\tau(p)$ from $F(x; 0)$ and (2.100). Finally, the desired solution $u(x, t)$ is constructed using (2.13), (2.81), and (2.5). This procedure is illustrated in Fig. 2.1.

2.2.3 *Periodic Wave Solution*

The bilinear transformation method discussed in this chapter is also used to obtain the periodic wave solutions of some nonlinear evolution equations. We now illustrate the method of solution developed by Nakamura [48, 49] for the case of the KdV equation.

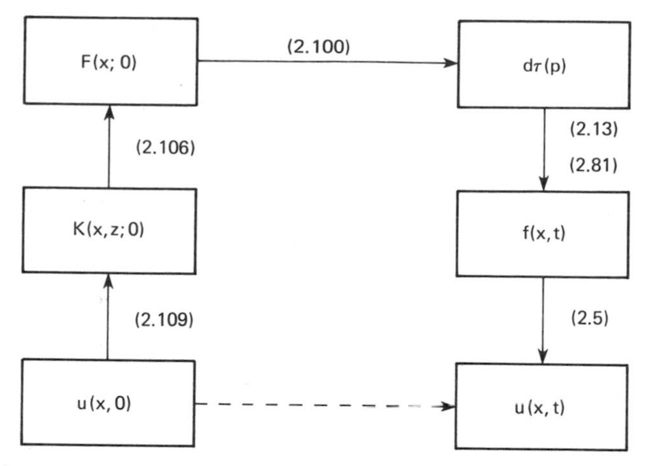

Fig. 2.1 Procedure for solving the initial value problem of the KdV equation.

The bilinear form of the KdV equation appropriate to the periodic problem is written as

$$F(D_t, D_x)f \cdot f = (D_t D_x + 6u_0 D_x^2 + D_x^4 + c)f \cdot f = 0, \quad (2.110)$$

with the dependent variable transformation

$$u(x, t) = u_0 + 2\,\partial^2 \ln f(x, t)/\partial x^2, \quad (2.111)$$

where u_0 is a constant and c an integration constant generally dependent on time.

For a one-periodic wave solution, we take the one-dimensional Riemann theta function in the form

$$f = \vartheta(\eta; \tau) = \sum_{n=-\infty}^{\infty} \exp(2\pi i n\eta + \pi i n^2 \tau), \qquad i = \sqrt{-1}, \quad (2.112)$$

with

$$\eta = px + \Omega t + \eta_0, \quad (2.113)$$

where τ is a complex constant satisfying the condition

$$\operatorname{Im} \tau > 0 \quad (2.114)$$

and p and Ω represent the wave number and frequency, respectively, and η_0 is a phase constant. Substituting (2.112) into (2.110) and using formula (App. I.2.1), we obtain

$$
\begin{aligned}
Ff\cdot f &= \sum_{n,n'=-\infty}^{\infty} F(D_t, D_x)\exp(2\pi in\eta + \pi in^2\tau)\cdot\exp(2\pi in'\eta + \pi in'^2\tau) \\
&= \sum_{n,n'=-\infty}^{\infty} F[2\pi i(n-n')\Omega, 2\pi i(n-n')p] \\
&\quad \times \exp[2\pi i(n+n')\eta + \pi i(n^2+n'^2)\tau] \\
&= \sum_{n,m=-\infty}^{\infty} F[2\pi i(2n-m)\Omega, 2\pi i(2n-m)p] \\
&\quad \times \exp\{2\pi im\eta + \pi i[n^2+(n-m)^2]\tau\} \\
&= \sum_{m=-\infty}^{\infty} \tilde{F}(m)\exp(2\pi im\eta),
\end{aligned} \tag{2.115}
$$

where the new summation index $m = n + n'$ has been introduced and $\tilde{F}(m)$ is defined by

$$
\begin{aligned}
\tilde{F}(m) = \sum_{n=-\infty}^{\infty} & F[2\pi i(2n-m)\Omega, 2\pi i(2n-m)p] \\
& \times \exp\{\pi i[n^2+(n-m)^2]\tau\}.
\end{aligned} \tag{2.116}
$$

Shifting index n by unity as $n = n' + 1$, (2.116) becomes

$$
\begin{aligned}
\tilde{F}(m) &= \sum_{n'=-\infty}^{\infty} F\{2\pi i[2n'-(m-2)]\Omega, 2\pi i[2n'-(m-2)]p\} \\
&\quad \times \exp\{\pi i[n'^2+(m-2-n')^2]\tau\}\exp[2\pi i(m-1)\tau] \\
&= \tilde{F}(m-2)\exp[2\pi i(m-1)\tau].
\end{aligned} \tag{2.117}
$$

From relation (2.117) we can conclude that, if $\tilde{F}(0)$ and $\tilde{F}(1)$ are zero, then all $\tilde{F}$'s become zero. This implies that (2.112) is an exact solution of the bilinear equation (2.110). Examining the definition of $\tilde{F}(m)$, the conditions $\tilde{F}(0) = \tilde{F}(1) = 0$ become

$$
\sum_{n=-\infty}^{\infty} F(4\pi in\Omega, 4\pi inp)\exp(2\pi in^2\tau) = 0, \tag{2.118}
$$

$$
\sum_{n=-\infty}^{\infty} F[2\pi i(2n-1)\Omega, 2\pi i(2n-1)p]\exp\{\pi i[n^2+(n-1)^2]\} = 0. \tag{2.119}
$$

Substituting expression (2.110) for F, (2.118) and (2.119) reduce to

$$\sum_{n=-\infty}^{\infty} (-16\pi^2 n^2 p\Omega - 96\pi^2 n^2 p^2 u_0 + 256\pi^4 n^4 p^4 + c) \exp(2\pi i n^2 \tau) = 0, \tag{2.120}$$

$$\sum_{n=-\infty}^{\infty} [-4\pi^2(2n-1)^2 p\Omega - 24\pi^2(2n-1)^2 u_0 + 16\pi^4(2n-1)^4 p^4 + c] \times \exp\{\pi i[n^2 + (n-1)^2]\} = 0, \tag{2.121}$$

respectively. Introducing the quantities

$$A_0(\tau) = \sum_{n=-\infty}^{\infty} \exp(2\pi i n^2 \tau), \tag{2.122}$$

$$A_1(\tau) = \sum_{n=-\infty}^{\infty} (4n)^2 \exp(2\pi i n^2 \tau), \tag{2.123}$$

$$A_2(\tau) = \sum_{n=-\infty}^{\infty} (4n)^4 \exp(2\pi i n^2 \tau), \tag{2.124}$$

$$B_0(\tau) = \sum_{n=-\infty}^{\infty} \exp\{\pi i[n^2 + (n-1)^2]\tau\}, \tag{2.125}$$

$$B_1(\tau) = \sum_{n=-\infty}^{\infty} (4n-2)^2 \exp\{\pi i[n^2 + (n-1)^2]\tau\}, \tag{2.126}$$

$$B_2(\tau) = \sum_{n=-\infty}^{\infty} (4n-2)^4 \exp\{\pi i[n^2 + (n-1)^2]\tau\}, \tag{2.127}$$

(2.120) and (2.121) can be written compactly as

$$-\pi^2 A_1 p\Omega - \pi^2 u_0 A_1 p^2 + \pi^4 A_2 p^4 + cA_0 = 0, \tag{2.128}$$

$$-\pi^2 B_1 p\Omega - \pi^2 u_0 B_1 p^2 + \pi^4 B_2 p^4 + cB_0 = 0. \tag{2.129}$$

From these two equations, quantities Ω and c are determined to be[§]

$$\Omega = -u_0 p + \pi^2 \frac{A_2 B_0 - A_0 B_2}{A_1 B_0 - A_0 B_1} p^3, \tag{2.130}$$

$$c = \frac{A_1 B_2 - A_2 B_1}{A_0 B_1 - A_1 B_0} (\pi p)^4. \tag{2.131}$$

[§] The important role of integration constant constant c, which has been taken to be zero in previous discussions of both soliton and generalized soliton solutions, is now evident. If we set $c = 0$ in (2.128) and (2.129), these two equations are obviously incompatible.

Expression (2.112), with (2.113) and (2.130), gives the one-periodic wave solution of the KdV equation. The soliton solution can be obtained from the periodic wave solution in an appropriate limiting procedure. To illustrate this we introduce a quantity

$$q = e^{\pi i \tau} \tag{2.132}$$

and take a limit $q \to 0$ (or $\operatorname{Im} \tau \to \infty$). The quantities defined in (2.122)–(2.127) are then expanded in powers of q as

$$A_0 = 1 + 2q^2 + 2q^3 + \cdots, \tag{2.133}$$

$$A_1 = 32q^2 + 128q^8 + \cdots, \tag{2.134}$$

$$A_2 = 512q^2 + 8192q^8 + \cdots, \tag{2.135}$$

$$B_0 = 2q + 2q^5 + 2q^{13} + \cdots, \tag{2.136}$$

$$B_1 = 8q + 72q^5 + 200q^{13} + \cdots, \tag{2.137}$$

$$B_2 = 32q + 2592q^5 + 20000q^{13} + \cdots. \tag{2.138}$$

Substituting these expressions into (2.130) and (2.131) yields

$$\begin{aligned} \Omega &= -u_0 p + 4\pi^2 \frac{1 - 30q^2 + 81q^4 + \cdots}{1 - 6q^2 + 9q^4 + \cdots} p^3 \\ &= -u_0 p + 4\pi^2 p^3, \qquad q \to 0, \end{aligned} \tag{2.139}$$

$$\begin{aligned} c &= -384q^2 \frac{1 - 15q^4 + 20q^6 + \cdots}{1 - 6q^2 + 9q^4 + \cdots} (\pi p)^4 \\ &= 0, \qquad q \to 0. \end{aligned} \tag{2.140}$$

By introducing the quantities

$$\tilde{p} = 2\pi i p, \tag{2.141}$$

$$\tilde{\Omega} = 2\pi i \Omega = -u_0 \tilde{p} - \tilde{p}^3, \tag{2.142}$$

$$\tilde{\eta}_0 = \eta_0 + \tau/2, \tag{2.143}$$

the one-periodic wave solution (2.112) reduces in the limit of $q \to 0$ (or $\operatorname{Im} \tau \to \infty$) to

$$\begin{aligned} f &= 1 + \exp(2\pi i \eta + \pi i \tau) + \exp(-2\pi i \eta + \pi i \tau) + \cdots \\ &= 1 + \exp(\tilde{p}x + \tilde{\Omega}t + \tilde{\eta}_0), \end{aligned} \tag{2.144}$$

which is simply the one-soliton solution of the KdV equation expressed in the bilinear variable.

The extension of these results to the N-periodic wave solution is straightforward. The solution is expressed in terms of the multidimensional theta function as

$$f(x, t) = \vartheta(\eta_1, \eta_2, \ldots, \eta_N; \tau) = \sum_{n_1, \ldots, n_N = -\infty}^{\infty} \exp\left(2\pi i \sum_{j=1}^{N} n_j \eta_j + \pi i \sum_{j,k=1}^{N} \tau_{jk} n_j n_k\right), \quad (2.145)$$

with

$$\eta_j = p_j x + \Omega_j t + \eta_{0j}, \qquad j = 1, 2, \ldots, N, \quad (2.146)$$

and p_j, Ω_j, and η_{0j} defined as in the one-periodic wave case. The term τ_{jk} $(j \neq k)$ represents the effect of interaction between periodic waves and is assumed to satisfy the conditions

$$\tau_{jk} = \tau_{kj}, \qquad j, k = 1, 2, \ldots, N, \quad (2.147)$$

$$\operatorname{Im} \tau_{jk} > 0, \qquad j, k = 1, 2, \ldots, N. \quad (2.148)$$

The relation corresponding to (2.115) is now written as

$$Ff \cdot f = \sum_{m_1, \ldots, m_N = -\infty}^{\infty} \tilde{F}(m_1, \ldots, m_N) \exp\left(2\pi i \sum_{j=1}^{N} m_j \eta_j\right), \quad (2.149)$$

where

$$\begin{aligned}\tilde{F}(m_1, \ldots, m_N) &= \sum_{n_1, \ldots, n_N = -\infty}^{\infty} F\left[2\pi i \sum_{j=1}^{N} (2n_j - m_j)\Omega_j, 2\pi i \sum_{j=1}^{N} (2n_j - m_j)p_j\right] \\ &\quad \times \exp\left\{\pi i \sum_{j,k=1}^{N} [n_j \tau_{jk} n_k + (m_j - n_j)\tau_{jk}(m_k - n_k)]\right\}. \quad (2.150)\end{aligned}$$

By shifting the hth summation index n_h by unity, we obtain the relation corresponding to (2.117):

$$\begin{aligned}\tilde{F}(m_1, \ldots, m_N) &= \tilde{F}(m_1, \ldots, m_{h-1}, m_h - 2, m_{h+1}, \ldots, m_N) \\ &\quad \times \exp\left(2\pi i \sum_{j=1}^{N} \tau_{hj} m_j - 2\tau_{hh}\right), \qquad h = 1, 2, \ldots, N. \quad (2.151)\end{aligned}$$

If relations

$$\tilde{F}(m_1, \ldots, m_N) = 0 \quad (2.152)$$

hold for all combinations of $m_1 = 0, 1, m_2 = 0, 1, \ldots, m_N = 0, 1$, expression (2.145) with (2.146) gives the N-periodic wave solutions of the KdV equation. Whether relations (2.152) are sufficient to determine the unknowns depends on the explicit functional form of $F(D_t, D_x)$. We can at least say that the unknown quantities included in the problem are frequencies Ω_j ($j = 1, 2, \ldots, N$), interaction terms τ_{jk} ($j \neq k$); $j, k = 1, 2, \ldots, N$), and an integration constant c, the total number of these being

$$N + {}_N C_2 + 1 = \tfrac{1}{2}(N^2 + N + 2). \tag{2.153}$$

The number of equations is given by 2^N [from (2.151)]. For $N = 1, 2$, the number of equations corresponds to that of unknowns, implying the existence of the one- and two-periodic wave solutions. For further details, including a discussion of the extension of these results to other types of nonlinear evolution equations such as the modified KdV equation and the Kadomtsev–Petviashvili (or the two-dimensional KdV) equation, refer to the original publications by Nakamura [48, 49].

2.3 Bäcklund Transformation

The Bäcklund transformation is a transformation that relates pairs of solutions of nonlinear evolution equations. It is a powerful method for analyzing some classes of nonlinear evolution equations. The Bäcklund transformation is used to obtain an N-soliton solution by purely algebraic means. It also provides a method for constructing an infinite number of conservation laws in a systematic way.[§] In this section we shall derive the Bäcklund transformation of the KdV equation on the basis of the bilinear transformation method [13].

Let f and f' be two solutions of the bilinearized KdV equation (2.8), that is,

$$D_x(D_t + D_x^3) f \cdot f = 0, \tag{2.154}$$

$$D_x(D_t + D_x^3) f' \cdot f' = 0. \tag{2.155}$$

and consider an equation

$$[D_x(D_t + D_x^3) f' \cdot f'] ff - f'f'[D_x(D_t + D_x^3) f \cdot f] = 0. \tag{2.156}$$

[§] An introductory monograph which treats the Bäcklund transformations of nonlinear evolution equations that appear in physics and applied mathematics has been written by Rogers and Shadwick [12].

Obviously, if f satisfies (2.154), f' also satisfies the same equation. Therefore, (2.156) can be regarded as a relation that connects the pair of solutions f and f', that is, the Bäcklund transformation of the KdV equation. Using formulas (App. I.6.2) and (App. I.6.5), (2.156) is converted to

$$\begin{aligned} &2D_x[(D_t + 3\lambda D_x + D_x^3)f' \cdot f] \cdot (ff') \\ &\quad + 6D_x[(D_x^2 - \lambda)f' \cdot f] \cdot (D_x f \cdot f') = 0, \end{aligned} \tag{2.157}$$

where an arbitrary parameter λ has been introduced. Equation (2.157) is satisfied provided that the following equations hold:

$$(D_t + 3\lambda D_x + D_x^3)f' \cdot f = 0, \tag{2.158}$$

$$D_x^2 f' \cdot f = \lambda f'f. \tag{2.159}$$

These constitute the Bäcklund transformation of the KdV equation in the bilinear formalism. To see the relation between the present bilinear formalism and the Wahlquist and Estabrook formalism [50], we introduce functions w and w' as

$$w = 2\,\partial \ln f/\partial x, \tag{2.160}$$

$$w' = 2\,\partial \ln f'/\partial x. \tag{2.161}$$

From formula (App. I.7.2), with $a = b = f$ and $c = d = f'$,

$$\begin{aligned} &(D_x^2 f' \cdot f')ff + f'f'(D_x^2 f \cdot f) \\ &\quad = (D_x^2 f' \cdot f)ff' + f'f(D_x^2 f \cdot f') - 2(D_x f' \cdot f)(D_x f' \cdot f). \end{aligned} \tag{2.162}$$

Using (2.159)–(2.161), we then have

$$w'_x + w_x = 2\lambda - \tfrac{1}{2}(w' - w)^2. \tag{2.163}$$

Note that w' and w satisfy the nonlinear evolution equations

$$w'_t + 3(w'_x)^2 + w'_{xxx} = 0, \tag{2.164}$$

$$w_t + 3(w_x)^2 + w_{xxx} = 0, \tag{2.165}$$

respectively.

It follows from (2.158) and formulas (App. I.3.1)–(App. I. 3.3) that

$$\begin{aligned} &w'_t - w_t + 3\lambda(w'_x - w_x) + w'_{xxx} - w_{xxx} \\ &\quad + \tfrac{3}{2}[(w' - w)(w'_x + w_x)]_x + \tfrac{1}{4}[(w' - w)^3]_x = 0. \end{aligned} \tag{2.166}$$

Equations (2.163) and (2.166) correspond to the Bäcklund transformation first found by Wahlquist and Estabrook [50]. The Bäcklund transformations of the higher-order KdV equations will be derived in Section 5.5.

To illustrate the process of generating soliton solutions starting from a known solution, we shall take the simplest case

$$f = 1, \tag{2.167}$$

which is transformed into the original variable u as

$$u = 2\,\partial^2 \ln f/\partial x^2 = 0, \tag{2.168}$$

so that (2.167) can be regarded as a *vacuum* solution. Substitution of (2.167) into (2.158) and (2.159) yields the linear equations for f and f' as

$$f'_t + 3\lambda f'_x + f'_{xxx} = 0, \tag{2.169}$$

$$f'_{xx} = \lambda f'. \tag{2.170}$$

A solution that satisfies both (2.169) and (2.170) is

$$f' = e^{\eta_1/2} + e^{-\eta_1/2} = e^{-\eta_1/2}(1 + e^{\eta_1}), \tag{2.171}$$

with

$$\lambda = k_1^2/4, \tag{2.172}$$

where

$$\eta_j = p_j x - p_j^3 t + \eta_{0j}, \qquad j = 1, 2, \ldots. \tag{2.173}$$

Noting that

$$\partial^2 \ln e^{-\eta_1/2}/\partial x^2 = 0, \tag{2.174}$$

it can be seen that (2.171) gives a one-soliton solution of the KdV equation

$$\begin{aligned} 2\,\partial^2 \ln f'/\partial x^2 &= 2\,\partial^2 \ln(1 + e^{\eta_1})/\partial x^2 \\ &= 2\,\partial^2 \ln f_1/\partial x^2 = u_1. \end{aligned} \tag{2.175}$$

A two-soliton solution is generated from (2.158) and (2.159) with f' given by (2.171) and

$$\lambda = k_2^2/4, \tag{2.176}$$

yielding

$$\begin{aligned} f &= (p_1 - p_2)[e^{(\eta_1+\eta_2)/2} + e^{-(\eta_1+\eta_2)/2}] \\ &\quad - (p_1 + p_2)[e^{(\eta_1-\eta_2)/2} + e^{-(\eta_1-\eta_2)/2}] \\ &= (p_1 - p_2)e^{-(\eta_1+\eta_2)/2}\left[1 - \frac{p_1 + p_2}{p_1 - p_2}(e^{\eta_1} + e^{\eta_2}) + e^{(\eta_1+\eta_2)}\right]. \end{aligned} \tag{2.177}$$

Taking phase constants η_{01} and η_{02} as

$$\eta_{01} = \ln\frac{p_2 - p_1}{p_2 + p_1} + \tilde{\eta}_{01}, \tag{2.178}$$

$$\eta_{02} = \ln\frac{p_2 - p_1}{p_2 + p_1} + \tilde{\eta}_{02}, \tag{2.179}$$

with $\tilde{\eta}_{01}$ and $\tilde{\eta}_{02}$ being constants, it can be seen that (2.177) gives a two-soliton solution of the KdV equation [see (2.33)].

This procedure may be extended to generate multisoliton solutions. The relation that connects $(N-1)$-soliton, N-soliton, and $(N+1)$-soliton solutions is called the superposition formula and it will now be introduced by following the work of Hirota and Satsuma [51]. To derive this formula, the four solutions of (2.159) are taken to be

$$f_0 = 1, \tag{2.180}$$

$$f_1 = f_1(x, t; p_1), \tag{2.181}$$

$$f_2 = f_2(x, t; p_2), \tag{2.182}$$

$$f_{12} = f_{12}(x, t; p_1, p_2), \tag{2.183}$$

where f_1 and f_2 correspond to one-soliton solutions with parameters p_1 and p_2, respectively, and f_{12} to a two-soliton solution with parameters p_1 and p_2. These are satisfied by

$$(D_x^2 - p_1^2/4)f_0 \cdot f_1 = 0, \tag{2.184}$$

$$(D_x^2 - p_2^2/4)f_0 \cdot f_2 = 0, \tag{2.185}$$

$$(D_x^2 - p_1^2/4)f_2 \cdot f_{12} = 0, \tag{2.186}$$

$$(D_x^2 - p_2^2/4)f_1 \cdot f_{12} = 0. \tag{2.187}$$

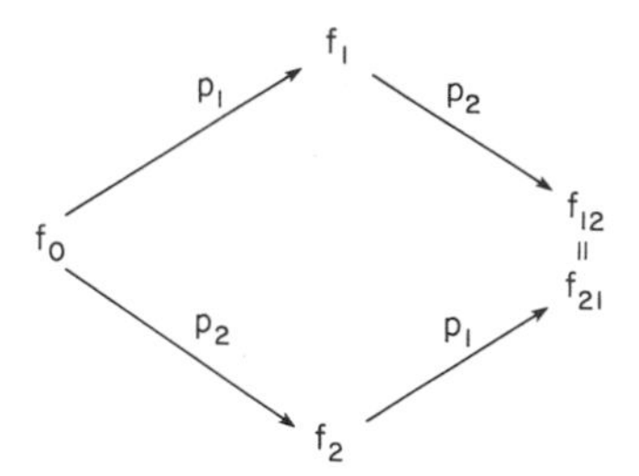

Fig. 2.2 Interrelationship among the four solutions f_0, f_1, f_2, and f_{12} $(=f_{21})$. See text for discussion.

In Fig. 2.2 the *commutability relation* $f_{12} = f_{21}$ has been assumed.[§]

We will now derive the superposition formula given these conditions. It follows from (2.184) and (2.186) that

$$(D_x^2 f_0 \cdot f_1) f_2 f_{12} - f_0 f_1 (D_x^2 f_2 \cdot f_{12}) = 0, \tag{2.188}$$

which is transformed, using formula (App. I.6.3), into the form

$$D_x[(D_x f_0 \cdot f_{12}) \cdot f_2 f_1 + f_0 f_{12} \cdot (D_x f_2 \cdot f_1)] = 0. \tag{2.189}$$

Using (2.185) and (2.187),

$$(D_x^2 f_0 \cdot f_2) f_1 f_{12} - f_0 f_2 (D_x^2 f_1 \cdot f_{12}) = 0, \tag{2.190}$$

which reduces to

$$D_x[(D_x f_1 \cdot f_2) \cdot f_0 f_{12} + f_1 f_2 \cdot (D_x f_0 \cdot f_{12})] = 0. \tag{2.191}$$

Adding (2.189) and (2.191) yields

$$D_x(f_0 f_{12}) \cdot D_x(f_1 \cdot f_2) = 0, \tag{2.192}$$

which gives the superposition formula [51]

$$f_0 f_{12} = c D_x(f_1 \cdot f_2) \tag{2.193}$$

by formula (App. I.1.5), where constant c is determined by the form of solutions. The subtraction of (2.191) from (2.189) yields

$$D_x[(D_x f_0 \cdot f_{12}) \cdot f_2 f_1] = 0, \tag{2.194}$$

[§] This relation will be proved later.

or equivalently,

$$f_1 f_2 = cD_x(f_0 \cdot f_{12}). \tag{2.195}$$

The generalization of (2.193) and (2.195) to $(N - 1)$-, N-, and $(N + 1)$-soliton solutions is straightforward, and the results may be expressed as

$$f_{N-1} f_{N+1} = cD_x(f_N \cdot \hat{f}_N), \tag{2.196}$$

$$f_N \hat{f}_N = cD_x(f_{N-1} \cdot f_{N+1}). \tag{2.197}$$

Here the parameter dependences of the four solutions f_{N-1}, f_N, $\hat{f}_N$, and f_{N+1} are written explicitly as

$$f_{N-1} = f_{N-1}(x, t; p_1, p_2, \ldots, p_{N-1}), \tag{2.198}$$

$$f_N = f_N(x, t; p_1, p_2, \ldots, p_N), \tag{2.199}$$

$$\hat{f}_N = \hat{f}_N(x, t; p_1, p_2, \ldots, p_{N-1}, p_{N+1}), \tag{2.200}$$

$$f_{N+1} = f_{N+1}(x, t; p_1, p_2, \ldots, p_{N-1}, p_N, p_{N+1}). \tag{2.201}$$

From these formulas, an N-soliton solution may be expressed in the form [51]

$$f_N = \sum_{\varepsilon=\pm 1} \frac{\prod_{j<k}^{(N)} (\varepsilon_j p_j - \varepsilon_k p_k)}{\prod_{j=1}^{N} \varepsilon_j p_j} \exp\left(\frac{1}{2} \sum_{j=1}^{N} \varepsilon_j \eta_j\right), \tag{2.202}$$

where $\sum_{\varepsilon=\pm 1}$ is the summation over all possible combinations of $\varepsilon_1 = \pm 1, \varepsilon_2 = \pm 1, \ldots, \varepsilon_N = \pm 1$.

The superposition formula (2.193) or (2.196) has been derived assuming the commutability relation $f_{12} = f_{21}$ (see Fig. 2.2). We shall now verify this relation by employing bilinear formalism. Define function $\tilde{f}_{12}$ as

$$\tilde{f}_{12} = cf_0^{-1} D_x f_1 \cdot f_2, \tag{2.203}$$

with $f_0 f_1 f_2 \neq 0$. It can be shown that $\tilde{f}_{12}$ is related by the Bäcklund transformation to f_1 and f_2, that is,

$$P_1 \equiv (D_x^2 - \tfrac{1}{4}p_2^2) f_1 \cdot \tilde{f}_{12} = 0, \tag{2.204}$$

$$P_2 \equiv (D_x^2 - \tfrac{1}{4}p_1^2) f_2 \cdot \tilde{f}_{12} = 0. \tag{2.205}$$

so if (2.184) and (2.185) are satisfied then (2.203) becomes a new solution.

To show (2.204) we use (2.184), (2.185), (App. I.8.2), and (2.203) to give

$$
\begin{aligned}
0 &= [(D_x^2 - \tfrac{1}{4}p_1^2)f_0 \cdot f_1]f_2 - [(D_x^2 - \tfrac{1}{4}p_2^2)f_0 \cdot f_2]f_1 \\
&= -2f_{0,x}(D_x f_1 \cdot f_2) + f_0(D_x f_1 \cdot f_2)_x + \tfrac{1}{4}(p_2^2 - p_1^2)f_0 f_1 f_2 \\
&= -2c^{-1}f_{0,x}f_0\tilde{f}_{12} + c^{-1}f_0(f_0\tilde{f}_{12})_x + \tfrac{1}{4}(p_2^2 - p_1^2)f_0 f_1 f_2 \\
&= f_0[-c^{-1}(f_{0,x}\tilde{f}_{12} - f_0\tilde{f}_{12,x}) + \tfrac{1}{4}(p_2^2 - p_1^2)f_1 f_2] \\
&= f_0[-c^{-1}D_x f_0 \cdot \tilde{f}_{12} + \tfrac{1}{4}(p_2^2 - p_1^2)f_1 f_2].
\end{aligned} \tag{2.206}
$$

Since $f_0 \neq 0$ we obtain from (2.206)

$$
D_x f_0 \cdot \tilde{f}_{12} = \tfrac{1}{4}c(p_2^2 - p_1^2)f_1 f_2. \tag{2.207}
$$

It follows that

$$
\begin{aligned}
P_1 f_0 f_2 &= [(D_x^2 - \tfrac{1}{4}p_2^2)f_1 \cdot \tilde{f}_{12}]f_0 f_2 - f_1\tilde{f}_{12}(D_x^2 - \tfrac{1}{4}p_2^2)f_0 \cdot f_2 \\
&= D_x[(D_x f_1 \cdot f_2) \cdot f_0\tilde{f}_{12} + f_1 f_2 \cdot (D_x f_0 \cdot \tilde{f}_{12})] = 0,
\end{aligned} \tag{2.208}
$$

where use has been made of (App. I.6.3), (App. I.1.5), (2.203), and (2.207). Noting (2.185) and $f_0 f_2 \neq 0$, we readily obtain (2.204). A similar argument leads to (2.205) and therefore the commutability relation $f_{12} = f_{21} = \tilde{f}_{12}$ has been proved.

So far we have been concerned only with the space part of the Bäcklund transformation. The commutability relation is also satisfied for the time part of the Bäcklund transformation, that is, if

$$
[D_t + \tfrac{1}{4}p_1^2 D_x + D_x^3]f_0 \cdot f_1 = 0, \tag{2.209}
$$

$$
[D_t + \tfrac{1}{4}p_2^2 D_x + D_x^3]f_0 \cdot f_2 = 0, \tag{2.210}
$$

then

$$
Q_1 \equiv [D_t + \tfrac{1}{4}p_2^2 D_x + D_x^3]f_1 \cdot \tilde{f}_{12} = 0, \tag{2.211}
$$

$$
Q_2 \equiv [D_t + \tfrac{1}{4}p_1^2 D_x + D_x^3]f_2 \cdot \tilde{f}_{12} = 0, \tag{2.212}
$$

where $\tilde{f}_{12}$ is defined by (2.203). Relations (2.211) and (2.212) imply the commutability relation $f_{12} = f_{21} = \tilde{f}_{12}$. The somewhat tedious proof of (2.211) and (2.212) can be carried out by using the formulas for the bilinear operators. To show (2.211), consider a quantity

$$
\begin{aligned}
Q_1 f_0 f_2 \equiv{}& [(D_t + \tfrac{3}{4}p_2^2 D_x + D_x^3)f_1 \cdot \tilde{f}_{12}]f_0 f_2 \\
&- f_1\tilde{f}_{12}(D_t + \tfrac{3}{4}p_2^2 D_x + D_x^3)f_0 \cdot f_2,
\end{aligned} \tag{2.213}
$$

where the second term on the right-hand side has been added since it is equal to zero by (2.210). Using (App. I.6.1) and (App. I.6.4), (2.213) becomes

$$Q_1 f_0 f_2 = (D_t + \tfrac{3}{4}p_2^2 D_x + \tfrac{1}{4}D_x^3) f_1 f_2 \cdot f_0 \tilde{f}_{12} + \tfrac{3}{4}D_x[(D_x^2 f_1 \cdot f_2)\cdot f_0 \tilde{f}_{12} + 2(D_x f_1 \cdot f_2)\cdot(D_x f_0 \cdot \tilde{f}_{12}) + f_1 f_2 \cdot (D_x^2 f_0 \cdot \tilde{f}_{12})]. \tag{2.214}$$

Substituting (2.203) and (2.207) into (2.214) and using (App. I.6.5) yields

$$Q_1 f_0 f_2 = c[D_t + \tfrac{3}{8}(p_1^2 + p_2^2)D_x + \tfrac{1}{2}D_x^3] f_1 f_2 \cdot (D_x f_1 \cdot f_2) + \tfrac{1}{4}cD_x(D_x^3 f_1 \cdot f_2)\cdot f_1 f_2 + \tfrac{3}{4}D_x f_1 f_2 \cdot (D_x^2 f_0 \cdot \tilde{f}_{12}). \tag{2.215}$$

Since f_1 and f_2 are solutions of the bilinearized KdV equations,

$$D_x(D_t + D_x^3) f_1 \cdot f_1 = 0, \tag{2.216}$$

$$D_x(D_t + D_x^3) f_2 \cdot f_2 = 0, \tag{2.217}$$

which give

$$[D_x(D_t + D_x^3) f_1 \cdot f_1] f_2 f_2 - f_1 f_1 D_x(D_t + D_x^3) f_2 \cdot f_2 = 0. \tag{2.218}$$

Using (App. I.6.5) and (App. I.6.2), (2.218) is transformed into

$$D_x^3(D_x f_1 \cdot f_2)\cdot f_1 f_2 = D_x f_1 f_2 \cdot (D_t f_1 \cdot f_2). \tag{2.219}$$

It follows by substituting (2.219) into the first term on the right-hand side of (2.215) that

$$Q_1 f_0 f_2 = \tfrac{1}{2}cD_x f_1 f_2 \cdot \{[D_t + \tfrac{3}{4}(p_1^2 + p_2^2)D_x - \tfrac{1}{2}D_x^3] f_1 \cdot f_2 + (3/2c)D_x^2 f_0 \cdot \tilde{f}_{12}\}. \tag{2.220}$$

However, from (2.209), (2.210), (2.184), and (2.185) we have

$$0 = [(D_t + \tfrac{3}{4}p_1^2 D_x + D_x^3) f_0 \cdot f_1] f_2 - [(D_t + \tfrac{3}{4}p_2^2 D_x + D_x^3) f_0 \cdot f_2] f_1 + 3[(D_x^2 - \tfrac{1}{4}p_1^2) f_0 \cdot f_1]_x f_2 - 3[(D_x^2 - \tfrac{1}{4}p_2^2) f_0 \cdot f_2]_x f_1. \tag{2.221}$$

Using formulas (App. I.8.1), (App. I.8.3), and (App. I.8.4) together with (2.203) and (2.207), (2.221) is converted to

$$\begin{aligned} 0 &= -f_0[D_t + \tfrac{3}{4}(p_1^2 + p_2^2)D_x - \tfrac{1}{2}D_x^3] f_1 \cdot f_2 + (3/c) f_0 (D_x f_0 \cdot \tilde{f}_{12})_x \\ &\quad - (6/c) f_{0,xx}(f_0 \tilde{f}_{12}) + (3/2c) f_0 (f_0 \tilde{f}_{12})_{xx} \\ &= -f_0[D_t + \tfrac{3}{4}(p_1^2 + p_2^2)D_x - \tfrac{1}{2}D_x^3] f_1 \cdot f_2 \\ &\quad + (3/c)(-\tfrac{1}{2} f_{0,xx}\tilde{f}_{12} + f_{0,x}\tilde{f}_{12,x} - \tfrac{1}{2} f_0 \tilde{f}_{12,xx}) f_0 \\ &= -f_0\{[D_t + \tfrac{3}{4}(p_1^2 + p_2^2)D_x - \tfrac{1}{2}D_x^3] f_1 \cdot f_2 + (3/2c)D_x^2 f_0 \cdot \tilde{f}_{12}\}. \end{aligned} \tag{2.222}$$

Comparing (2.222) with (2.220) shows

$$Q_1 f_0 f_2 = 0, \tag{2.223}$$

or

$$Q_1 = 0, \tag{2.224}$$

since $f_0 f_2 \neq 0$. Hence (2.211) has been verified. A similar argument leads to the proof of (2.212, and therefore we have completed the proof of the commutability relation for the time part of the Bäcklund transformation.

Finally, we shall consider an important structure of the Bäcklund transformation. We have seen that the Bäcklund transformation enables us to obtain new solutions starting with a known solution, such as a vacuum solution. It also provides a new nonlinear evolution equation which usually has a structure similar to the original one. In the case of the KdV equation, this new equation is derived from (2.158) and (2.159) as follows: Define a new dependent variable v by

$$v = i\frac{\partial}{\partial x}\ln\frac{f'}{f}. \tag{2.225}$$

Then (2.158) and (2.159) reduce to a single equation

$$v_t + 6\lambda v_x + 6v^2 v_x + v_{xxx} = 0, \tag{2.226}$$

using formulas (App. I.3.1), (App. I.3.2), and (App. I.3.3). Equation (2.226) is the modified KdV equation. A new nonlinear evolution equation, which we call the second modified KdV equation, is generated from the Bäcklund transformation of the modified KdV equation when the procedure used in the case of the KdV equation is repeated. Nakamura and Hirota [52] derived the multisoliton solution for the second modified KdV equation. Nakamura [53] generalized this procedure and derived the third modified KdV equation generated from the Bäcklund transformation of the second modified KdV equation. Nakamura also conjectured a procedure to obtain the nth ($n \geq 4$) modified KdV equation by employing the bilinear transformation method and clarified the structure of the infinite chain process of the Bäcklund transformation of the KdV equation.

2.4 Conservation Laws

Some class of nonlinear evolution equations is characterized by the existence of an infinite number of conservation laws. A conservation law associated with a nonlinear evolution equation such as the KdV equation is expressed in the form

$$T_x + X_x = 0, \tag{2.227}$$

where T is the conserved density and X the flux of T. If T is a polynomial in u and its x derivatives, it is called a *polynomial-conserved density*. Of course, T may include a nonlocal quantity such as the Hilbert transform (see Chapter 3). Integrating (2.227) with respect to x from $-\infty$ to ∞ and setting the boundary condition $X \to 0$ as $|x| \to \infty$ yields

$$\frac{d}{dt}\int_{-\infty}^{\infty} T\, dx = 0, \tag{2.228}$$

which implies that $\int_{-\infty}^{\infty} T\, dx$ is a constant of motion.

An infinite number of polynomial conservation laws will be derived starting with the Bäcklund transformation of the KdV equation presented in Section 2.3. A procedure used to derive an infinite number of conservation laws is as follows: Define a function W as

$$W = w' - w, \tag{2.229}$$

and substitute (2.229) into (2.163) and (2.166) to obtain

$$W_x + 2u = 2\lambda - \tfrac{1}{2}W^2, \tag{2.230}$$

$$W_t + 3\lambda W_x + W_{xxx} + \tfrac{3}{2}[W(W_x + 2u)]_x + \tfrac{1}{4}(W^3)_x = 0, \tag{2.231}$$

where we have used a relation

$$w_x = 2\, \partial^2 \ln f/\partial x^2 = u. \tag{2.232}$$

Using (2.230), (2.231) may be rewritten as

$$W_t + 6\lambda W_x + W_{xxx} - \tfrac{3}{2}W^2 W_x = 0. \tag{2.233}$$

Equation (2.231) has the form of (2.227), therefore W is a conserved density. To derive conservation laws explicitly, expand W as

$$W = 2\eta + \sum_{m=1}^{\infty} f_m \eta^{-m}, \qquad \eta = \sqrt{\lambda}, \tag{2.234}$$

substitute (2.234) into (2.230), and then compare the η^{-m} terms on both sides of (2.230). The result is expressed in the form of a recursion formula as

$$f_1 = -u, \tag{2.235}$$

$$f_{m+1} = -\frac{1}{2}f_{m,x} - \frac{1}{4}\sum_{s=1}^{m-1} f_s f_{m-s}, \qquad m \geq 1. \tag{2.236}$$

The first few f_n are

$$f_2 = \tfrac{1}{2}u_x, \tag{2.237}$$

$$f_3 = -\tfrac{1}{4}u_{xx} - \tfrac{1}{4}u^2, \tag{2.238}$$

$$f_4 = \tfrac{1}{8}u_{xxx} + \tfrac{1}{2}uu_x, \tag{2.239}$$

$$f_5 = -\tfrac{1}{16}u_{xxxx} - \tfrac{3}{8}uu_{xx} - \tfrac{5}{16}(u_x)^2 - \tfrac{1}{8}u^3. \tag{2.240}$$

It may be seen from these expressions that f_{2m} ($m = 1, 2, \ldots$) vanish when integrated with respect to x from $-\infty$ to ∞ and that only odd terms f_{2m+1} yield meaningful results. Equation (2.230) is the well-known Riccati equation and may be linearized by an appropriate transformation. Introducing a function R as

$$W = -(1 - R_x)/R \tag{2.241}$$

and substituting (2.241) into (2.230) yield

$$-2RR_{xx} + R_x^2 - 4(u - \lambda)R^2 = 1. \tag{2.242}$$

Differentiating (2.242) with respect to x and dividing by $-2R$, we obtain a linear equation for R as

$$R_{xxx} + 4(u - \lambda)R_x + 2u_x R = 0. \tag{2.243}$$

It may be verified from (2.233) and (2.241) that R is also a conserved density, that is,[§]

$$\frac{d}{dt}\int_{-\infty}^{\infty} R(x, t)\, dx = 0. \tag{2.244}$$

[§] Note the relation

$$\begin{aligned} &W_t + 6\lambda W_x + W_{xxx} - \tfrac{3}{2}W^2 W_x \\ &\quad = (1/R^2)(1 - R_x + R\,\partial/\partial x)[R_t + 6\lambda R_x + R_{xxx} - \tfrac{3}{2}(R_x^2/R)_x + \tfrac{3}{2}(1/R)_x]. \end{aligned}$$

Expanding R in inverse powers of λ as

$$R = \sum_{m=0}^{\infty} R_m/\lambda^{m+1/2}, \qquad R_0 = -\tfrac{1}{2} \tag{2.245}$$

and substituting (2.245) into (2.243), we obtain a recursive formula for R_m

$$R_{m+1,x} = \tfrac{1}{4}(R_{m,xxx} + 4uR_{m,x} + 2u_x R_m), \qquad m \geq 0. \tag{2.246}$$

Formula (2.246) includes all the information about conservation laws and is very important to the study of the properties of conservation laws. Further details are related by Gel'fand and Dikii [54].

2.5 Inverse Scattering Method

In this section the inverse scattering formalism of the KdV equation, first developed by Gardner *et al.* [1], will be derived from the Bäcklund transformation expressed in terms of the bilinear operators.

Following Hirota [13], we introduce wave function ψ by the relation

$$f' = \psi f, \tag{2.247}$$

where f and f' satisfy (2.158) and (2.159). We then divide (2.158) by $f'f$ and use formulas (App. I.3.1)–(App. I.3.3) to obtain

$$-(\ln\psi)_t + 3\lambda(\ln\psi)_x + (\ln\psi)_{xx} + 3(\ln\psi)_x[(\ln\psi)_{xx} + 2(\ln f)_{xx}] + [(\ln\psi)_x]^3 = 0, \tag{2.248}$$

where we have used the identity

$$(\ln f'f)_{xx} = (\ln f'f^2/f)_{xx} = (\ln\psi)_{xx} + 2(\ln f)_{xx}. \tag{2.249}$$

Differentiating with respect to t and x and using (2.5), that is, $u = 2(\ln f)_{xx}$, we obtain the time evolution of ψ as

$$\psi_t + 3(u+\lambda)\psi_x + \psi_{xxx} = 0. \tag{2.250}$$

The space evolution of ψ is derived similarly from (2.159) using (App. I.3.2) and (2.5) as

$$(\ln\psi)_{xx} + 2(\ln f)_{xx} + [(\ln\psi)_x]^2 = \lambda, \tag{2.251a}$$

or

$$\psi_{xx} + u\psi = \lambda\psi. \tag{2.251b}$$

Substituting $\lambda\psi$ from (2.251b) into (2.250) yields another expression of the time evolution of ψ as

$$\psi_t = -4\psi_{xxx} - 6u\psi_x - 3u_x\psi. \qquad (2.252)$$

Equations (2.251) and (2.252) are the basis for the inverse scattering transform of the KdV equation [1].[§]

2.6 Bibliography

The bilinear transformation method has been illustrated by applying it to the KdV equation, a typical nonlinear evolution equation. From the bilinearized KdV equation various exact solutions, the Bäcklund transformations, an infinite number of conservation laws, and the inverse scattering transform have been derived in a systematic way; their interrelation is shown in Fig. 1.1. We believe that the essential part of the bilinear transformation method has been fully discussed. However, other references describe the applications of this method to other nonlinear evolution equations. Therefore, we shall describe some of the references related to the bilinear transformation method.

The original concept of the bilinear transformation method was described by Hirota [9], who used the bilinearized KdV equation (2.6) to derive an N-soliton solution of the KdV equation. The method has been applied to other nonlinear evolution equations, including the modified KdV equation [55]

$$v_t + 6v^2 v_x + v_{xxx} = 0; \qquad (2.253)$$

the Sine–Gordon equation [56]

$$v_{xt} = \sin v; \qquad (2.254)$$

a nonlinear wave equation with envelope-soliton solutions [57]

$$i\psi_t + i3\alpha|\psi|^2\psi_x + \beta\psi_{xx} + i\gamma\psi_{xxx} + \delta|\psi|^2\psi = 0, \qquad (2.255)$$

[§] The method of solution using (2.251) and (2.252) has been fully discussed [43] and we shall not go into detail here. Refer to textbooks listed in the references [5–8] for details of the inverse scattering method.

where α, β, γ, and δ are real constants which satisfy a relation $\alpha\beta = \gamma\delta$; the Boussinesq equation [58]

$$u_{tt} - u_{xxx} - 3(u^2)_{xx} - u_{xxxx} = 0; \tag{2.256}$$

the model equation for shallow-water waves [59]

$$u_t - u_{xxt} - 3uu_t + 3u_x \int_x^\infty u_t \, dx' + u_x = 0; \tag{2.257}$$

the cylindrical KdV equation [60, 61]

$$u_t + 6uu_x + u_{xxx} + u/2t = 0; \tag{2.258}$$

and the derivative nonlinear Schrödinger equation [62]

$$iu_t + \beta u_{xx} + i\delta' u^* u u_x + \delta u^* u u = 0, \tag{2.259}$$

where β, δ', and δ are real constants and * denotes complex conjugate.

The bilinear transformation method has also been applied to obtain exact solutions for certain classes of nonlinear *integrodifferential* equations. The first example is the Benjamin–Ono (BO) equation

$$u_t + 4uu_x + Hu_{xx} = 0, \tag{2.260}$$

where H is the Hilbert transform operator defined by

$$Hu(x, t) = \frac{1}{\pi} P \int_{-\infty}^{\infty} \frac{u(y, t)}{y - x} \, dy. \tag{2.261}$$

The first bilinearization of (2.260) and the N-soliton solution using the bilinear transformation method was described in Ref. [63]. Additionally, the bilinearized BO equation was used to obtain the periodic wave solution in Ref. [64]. The mathematical structure of the BO equation will be studied in Chapter 3.

The second example is the finite-depth fluid equation

$$u_t + 2uu_x + Gu_{xx} = 0, \tag{2.262}$$

with

$$Gu(x, t) = (1/2d)P \int_{-\infty}^{\infty} [\coth \pi(y - x)/2d - \operatorname{sgn}(y - x)]u(y, t) \, dy, \tag{2.263}$$

where d is the fluid depth. It should be noted that (2.262) reduces to the BO equation in the deep-water limit $d \to \infty$ and to the KdV equation in the shallow-water limit $d \to 0$. The bilinearization of (2.261) and the N-soliton solution were given in Refs. [27, 65]. For the development of bilinearization of soliton equations the reader is referred to Ref. [39].

These equations are concerned with the one-dimensional system. The two-dimensional nonlinear evolution equations can be treated similarly. The two-dimensional Sine–Gordon equation [66]

$$v_{xx} + v_{yy} - v_{tt} = \sin v, \tag{2.264}$$

the two-dimensional KdV (or Kadomtsev–Petviashvili) equation [67]

$$u_{tx} + 12(uu_x)_x + u_{xxxx} \pm 12u_{yy} = 0, \tag{2.265}$$

and the two-dimensional nonlinear Schrödinger equation [68]

$$iu_t + \beta u_{xx} + \beta' u_{yy} + \delta u^* uu = 0, \tag{2.266}$$

with β, β', and δ being real constants, are typical examples.

Other important classes of nonlinear evolution equations to which the bilinear transformation method has been successfully applied are nonlinear differential-difference and nonlinear partial difference equations. A famous example is the Toda equation [69]

$$d^2 \ln(1 + V_n)/dt^2 = V_{n-1} + V_{n+1} - 2V_n, \tag{2.267}$$

which was discussed in detail in Ref. [70]. The self-dual nonlinear network equation [71]

$$\frac{1}{1 + V_n^2} \frac{dV_n}{dt} = I_n - I_{n+1}, \tag{2.268}$$

$$\frac{1}{1 + I_n^2} \frac{dI_n}{dt} = V_{n-1} - V_n, \tag{2.269}$$

also belongs to a class of nonlinear differential-difference equations which can be bilinearized through the dependent variable transformation. Another type of nonlinear network equation describes a Volterra system

$$d \ln(z^{-1} + V_n)/dt = I_{n-1} - I_n, \tag{2.270}$$

$$d \ln(z^{-1} + I_n)/dt = V_n - V_{n+1}, \tag{2.271}$$

with z being a characteristic parameter of the network, was bilinearized and an N-soliton solution was presented in Ref. [72]. Nonlinear partial difference equations, where the difference analogues of the

KdV, Toda, Sine–Gordon, Liouville, two-wave interaction, Riccati, and Burgers equations are treated, were discussed in Refs. [73–77].

Hirota [78] proposed the discrete analogue of a generalized Toda equation

$$(z_1 e^{D_1} + z_2 e^{D_2} + z_3 e^{D_3}) f \cdot f = 0, \tag{2.272}$$

where z_i $(i = 1, 2, 3)$ are arbitrary constants and D_i $(i = 1, 2, 3)$ linear combinations of the bilinear operators D_t, D_x, D_y, D_n, etc., and showed that it reduces to various types of nonlinear evolution equations by appropriate choice of z_i and D_i.

A new formulation of the Bäcklund transformation was presented by Hirota using his bilinear transformation method [13]. The bilinear operators defined in the form of (2.7) were first introduced in the same paper. As noted in the last part of Section 2.3, new nonlinear evolution equations are generated from the Bäcklund transformation of a given nonlinear evolution equation. A variety of nonlinear network equations generated from the Bäcklund transformation for the Toda equation (2.267) were presented in Refs. [11, 79] together with their N-soliton solutions. The Boussinesq equation (2.256) [80], the KdV equation (2.1) [52, 53], the BO equation (2.260) [32], the finite-depth fluid equation, and the Sine–Gordon equation (2.254) [81, 82] were also employed to generate a new class of nonlinear evolution equations.

The Lax hierarchy of nonlinear evolution equations is also a very important class of nonlinear evolution equations. First introduced by Lax [2] in the study of the structure of the KdV equation, they are called the higher-order equations in this book. A systematic method for bilinearizing the higher-order KdV equations was developed by Matsuno [22] and was applied to the higher-order equations for the modified KdV equation (2.253) [30], the nonlinear Schrödinger equation (2.255) with $\alpha = \gamma = 0$ [30], the BO equation (2.260) [19–21], and the finite-depth fluid equation (2.263) [21,29]. The method of bilinearization for higher-order equations will be presented in Chapter 5.

Finally, we shall mention some references concerning various types of solutions of nonlinear evolution equations. The soliton and periodic wave solutions are typical of those that are nonsingular for time and space variables. The generalized soliton solutions discussed in Section 2.2.2 are combinations of soliton and ripple solutions, the latter being the form of dispersive waves. Along with these solutions, singular solutions also exist. The rational solutions of the KdV equation with decay $-2/x^2$ as $|x| \to \infty$ are typical. The bilinear transformation method was used to obtain rational solutions of the KdV equation [83, 84]. Another class of solutions, similarity-type decay-mode (or ripplon) solutions, also exist in two-dimensional nonlinear systems; the Bäcklund transformation in the bilinear formalism was applied to obtain ripplon solutions for the two-dimensional KdV equation [85, 86], the two-dimensional nonlinear Schrödinger equation [87], and the two-dimensional Toda equation [88].

3

The Benjamin–Ono Equation

This chapter is concerned with the mathematical structure of the Benjamin–Ono equation. Three different methods are presented for obtaining the N-soliton solution. First, the bilinear transformation method is used to obtain the N-soliton and the N-periodic wave solutions of the BO equation. Second, it is shown that the N-soliton solution is derived from the system of N linear algebraic equations. Third, the pole expansion method is applied to the BO equation to obtain the N-soliton solution.

The Bäcklund transformation of the BO equation is then constructed on the basis of the bilinear transformation method, and it is employed to derive an infinite number of conserved quantities and the inverse scattering transform of the BO equation. A method for solving the initial value problem of the BO equation is then developed, and the properties of solutions are investigated using a zero dispersion limit. Finally, the stability of the BO solitons and the linearized BO equation are briefly discussed.

3.1 Multisoliton Solutions of the Benjamin–Ono Equation

3.1.1 *Derivation of the N-Soliton Solution by the Bilinear Transformation Method*

The BO equation describes a large class of internal waves in a stratified fluid of great depth [15–17], and it also governs the propagation of nonlinear Rossby waves in a rotating fluid [18]. The BO equation may be written in the form

$$u_t + 4uu_x + Hu_{xx} = 0, \tag{3.1}$$

where H denotes the Hilbert transform operator defined by

$$Hu(x, t) = \frac{1}{\pi} P \int_{-\infty}^{\infty} \frac{u(y, t)}{y - x}\, dy. \tag{3.2}$$

The Hilbert transform has a dispersive effect in the BO equation. Note that the Hilbert transform is a definite integral, which differs from an integral term that appears in the model equation for shallow-water waves [59], and it makes the properties of these solutions very different from those of the well-known KdV type. Nevertheless, as will be shown in this chapter, the BO equation shares many of the properties of the KdV and related nonlinear partial differential equations.

In this section the N-soliton solution of the BO equation is derived by three different methods: the bilinear transformation method [63, 64]; the theory of linear algebraic equation [89]; and the pole expansion method [90, 91].

We shall now employ the bilinear transformation method discussed in Chapter 2. The one-soliton solution of (3.1) has a Lorentzian profile and is expressed as

$$u_s(x, t) = \frac{a}{a^2(x - at - x_0)^2 + 1}, \tag{3.3}$$

where a (>0) and x_0 are the amplitude and phase, respectively, of the soliton. To infer the form of a dependent variable transformation that enables us to transform (3.1) into a bilinear equation, we deform the one-soliton solution (3.3) as

$$\begin{aligned} u_s(x, t) &= \frac{i}{2}\left[\frac{a}{a(x - at - x_0) + i} - \frac{a}{a(x - at - x_0) - i}\right], \quad i = \sqrt{-1}, \\ &= \frac{i}{2}\frac{\partial}{\partial x}\ln\frac{a(x - at - x_0) + i}{a(x - at - x_0) - i} \\ &= \frac{i}{2}\frac{\partial}{\partial x}\ln\frac{-i(x - at - x_0) + 1/a}{i(x - at - x_0) + 1/a} \\ &= \frac{i}{2}\frac{\partial}{\partial x}\ln\frac{f_1^*}{f_1}, \end{aligned} \tag{3.4}$$

where

$$f_1 = i(x - at - x_0) + 1/a, \tag{3.5}$$

and * denotes the complex conjugate. For the N-soliton case, we can expect the solution to be represented by a superposition of the one-soliton solution (3.3) in the limit of large values of t. In this limit the N-soliton solution may be represented in the form

$$\begin{aligned} u(x, t) &\simeq \sum_{j=1}^{N} \frac{i}{2}\frac{\partial}{\partial x}\ln\frac{f_j^*}{f_j} \\ &= \frac{i}{2}\frac{\partial}{\partial x}\ln\left(\prod_{j=1}^{N} f_j^* \Big/ \prod_{j=1}^{N} f_j\right), \end{aligned} \tag{3.6}$$

where

$$f_j = i\theta_j + 1/a_j, \qquad j = 1, 2, \ldots, N, \tag{3.7}$$

with

$$\theta_j = x - a_j t - x_{0j}, \qquad j = 1, 2, \ldots, N. \tag{3.8}$$

Here a_j (>0) and x_{0j} are the amplitude and phase, respectively, of the jth soliton, and it is assumed that $a_j \neq a_k$ for $j \neq k$. Examining (3.6), we write the dependent variable transformation in the form

$$u(x, t) = \frac{i}{2} \frac{\partial}{\partial x} \ln \frac{f^*(x, t)}{f(x, t)}, \tag{3.9}$$

$$f(x, t) \propto \prod_{j=1}^{N} [x - x_j(t)], \tag{3.10}$$

$$\text{Im } x_j(t) > 0, \qquad j = 1, 2, \ldots, N, \tag{3.11}$$

where x_j $(j = 1, 2, \ldots, N)$ are complex functions of time t whose imaginary parts are positive.

Using the formulas

$$H[1/(x - x_j)] = -i/(x - x_j) \tag{3.12}$$

$$H[1/(x - x_j^*)] = i/(x - x_j^*), \tag{3.13}$$

which are consequences of (3.11), we obtain

$$\begin{aligned} Hu(x, t) &= \frac{i}{2} H \frac{\partial}{\partial x} \ln \frac{f^*}{f} \\ &= \frac{i}{2} H \sum_{j=1}^{N} \left(\frac{1}{x - x_j^*} - \frac{1}{x - x_j} \right) \\ &= \frac{i}{2} \sum_{j=1}^{N} \left(\frac{i}{x - x_j^*} + \frac{i}{x - x_j} \right) \\ &= -\frac{1}{2} \frac{\partial}{\partial x} \ln(f^* f). \end{aligned} \tag{3.14}$$

Substituting (3.9) and (3.14) into (3.1) yields

$$\frac{\partial}{\partial x} \left[\frac{i}{2} \frac{\partial}{\partial t} \ln \frac{f^*}{f} + 2\left(\frac{i}{2} \frac{\partial}{\partial x} \ln \frac{f^*}{f} \right)^2 - \frac{\partial^2}{\partial x^2} \ln(f^* f) \right] = 0. \tag{3.15}$$

Integrating (3.15) with respect to x and differentiating we obtain [63]

$$i(f_t^* f - f^* f_t) - (f_{xx}^* f - 2f_x^* f_x + f_{xx} f^*) = 0, \tag{3.16}$$

where an integration constant is assumed to be zero. Equation (3.16) can be rewritten in terms of the bilinear operators introduced in Chapter 2 as

$$iD_t f^* \cdot f = D_x^2 f^* \cdot f, \tag{3.17}$$

which is the bilinearized BO equation.

The one-soliton solution of (3.17) is given by (3.5), that is,

$$f_1 = i\theta_1 + 1/a_1, \tag{3.18}$$

which can be verified by direct substitution.

For the two-soliton solution, we assume f_2 is of the form

$$f_2 = c_3\theta_1\theta_2 + c_1\theta_1 + c_2\theta_2 + b, \tag{3.19}$$

where c_1, c_2, c_3, and b are unknown constants. Since (3.17) is invariant under a scaling $f \to cf$ (c constant), then constant c_3 in (3.19) may be arbitrarily chosen. Setting

$$c_3 = -1, \tag{3.20}$$

f_2 becomes

$$f_2 = -\theta_1\theta_2 + c_1\theta_2 + c_2\theta_2 + b. \tag{3.21}$$

It follows by direct calculation that

$$\begin{aligned} iD_t f_2^* \cdot f_2 = {} & i(c_1 - c_1^*)a_2\theta_1^2 + i(c_2 - c_2^*)a_1\theta_2^2 \\ & - ia_2(c_1c_2^* - c_1^*c_2 - b + b^*)\theta_1 \\ & - ia_1(c_2c_1^* - c_1c_2^* - b^* + b)\theta_2 \\ & - i(a_1c_1^* + a_2c_2^*)b + i(a_1c_1 + a_2c_2)b^* \end{aligned} \tag{3.22}$$

and

$$\begin{aligned} D_x^2 f_2^* \cdot f_2 = {} & -2[\theta_1^2 + \theta_2^2 + (c_2 + c_2^*)\theta_1 + (c_1 + c_1^*)\theta_2 \\ & + b + b^* + (c_1 + c_2)(c_1^* + c_2^*)]. \end{aligned} \tag{3.23}$$

By comparing $\theta_1^2, \theta_2^2, \theta_1, \theta_2$ and constant terms on both sides of (3.17), the equations required to determine the unknown constants c_1, c_2, and b may be derived. The equations

$$i(c_1 - c_1^*)a_2 = -2, \tag{3.24}$$

$$i(c_2 - c_2^*)a_1 = -2, \tag{3.25}$$

$$ia_2(c_1c_2^* - c_1^*c_2 - b + b^*) = 2(c_2 + c_2^*), \tag{3.26}$$

$$ia_1(c_1^*c_2 - c_1c_2^* - b^* + b) = 2(c_1 + c_1^*), \tag{3.27}$$

$$\begin{aligned} & -i(a_1c_1^* + a_2c_2^*)b + i(a_1c_1 + a_2c_2)b^* \\ & \quad = -2[b + b^* + (c_1 + c_2)(c_1^* + c_2^*)] \end{aligned} \tag{3.28}$$

are satisfied by the constants

$$c_1 = i/a_2, \tag{3.29}$$

$$c_2 = i/a_1, \tag{3.30}$$

$$b = \frac{1}{a_1 a_2}\left(\frac{a_1 + a_2}{a_1 - a_2}\right)^2, \tag{3.31}$$

which, substituted into (3.21), give the two-soliton solution

$$f_2 = -\theta_1\theta_2 + i\left(\frac{\theta_1}{a_2} + \frac{\theta_2}{a_1}\right) + \frac{1}{a_1 a_2}\left(\frac{a_1 + a_2}{a_1 - a_2}\right)^2. \tag{3.32}$$

Note that f_2 is expressed as the determinant

$$f_2 = \begin{vmatrix} i\theta_1 + 1/a_1 & 2/(a_1 - a_2) \\ 2/(a_2 - a_1) & i\theta_2 + 1/a_2 \end{vmatrix}. \tag{3.33}$$

Repeating the same procedure, the three-soliton solution f_3 is found explicitly as

$$\begin{aligned} f_3 = & -i\theta_1\theta_2\theta_3 - \left(\frac{\theta_1\theta_2}{a_3} + \frac{\theta_2\theta_3}{a_1} + \frac{\theta_3\theta_1}{a_2}\right) \\ & + i\left(\frac{B_{23}\theta_1}{a_2 a_3} + \frac{B_{31}\theta_2}{a_3 a_1} + \frac{B_{12}\theta_3}{a_1 a_2}\right) + \frac{B_{12} + B_{23} + B_{31} - 2}{a_1 a_2 a_3} \\ = & \begin{vmatrix} i\theta_1 + 1/a_1 & 2/(a_1 - a_2) & 2/(a_1 - a_3) \\ 2/(a_2 - a_1) & i\theta_2 + 1/a_2 & 2/(a_2 - a_3) \\ 2/(a_3 - a_1) & 2/(a_3 - a_2) & i\theta_3 + 1/a_3 \end{vmatrix}, \end{aligned} \tag{3.34}$$

where

$$B_{jk} = (a_j + a_k)^2/(a_j - a_k)^2, \qquad j, k = 1, 2, \ldots. \tag{3.35}$$

In general, the N-soliton solution f_N is given compactly as [63]

$$f_N = \det M, \tag{3.36}$$

where M is an $N \times N$ matrix whose elements are given by

$$M_{jk} = \begin{cases} i\theta_j + 1/a_j & \text{for } j = k, \quad (3.37) \\ 2/(a_j - a_k) & \text{for } j \neq k. \quad (3.38) \end{cases}$$

It may be confirmed by direct calculation that f_N given by (3.36) satisfies the bilinearized BO equation (3.17). However, this can be shown more easily by taking the long-wave limit of the N-periodic wave solution of the BO equation, as will be demonstrated later in this section.

It is now necessary to verify that the N-soliton solution (3.36) satisfies assumption (3.11) used in the process of deriving (3.17). We write the equation of motion of $x_n(t)$ in the form

$$\frac{d}{dt} x_n(t) = 2i\left(\sum_{\substack{s=1 \\ (s \neq n)}}^{N} \frac{1}{x_s - x_n} + \sum_{s=1}^{N} \frac{1}{x_n - x_s^*} \right), \qquad n = 1, 2, \ldots, N, \tag{3.39}$$

which is derived by substituting (3.9) with (3.10) into (3.1) and setting the coefficient of $[x - x_n(t)]^{-2}$ as zero [90, 91]. Taking the imaginary part of (3.39), we obtain the time evolution of the imaginary part of x_n as

$$d \operatorname{Im} x_n(t)/dt = G_n(t) \operatorname{Im} x_n(t), \qquad n = 1, 2, \ldots, N, \tag{3.40}$$

where

$$G_n(t) = 8 \sum_{\substack{s=1 \\ (s \neq n)}}^{N} \left(\frac{\operatorname{Re}(x_s - x_n) \operatorname{Im} x_s}{[\operatorname{Re}(x_s - x_n)]^2 + [\operatorname{Im}(x_s - x_n)]^2} \times \frac{1}{[\operatorname{Re}(x_s - x_n)]^2 + [\operatorname{Im}(x_s + x_n)]^2} \right). \tag{3.41}$$

Integration of (3.40) with respect to t yields

$$\operatorname{Im} x_n(t) = \operatorname{Im} x_n(t_0) \exp\left[\int_{t_0}^{t} G_n(t')\, dt' \right], \tag{3.42}$$

where t_0 is an initial time. It can be seen from (3.42) that conditions (3.11) are satisfied if they hold at some time t_0, since $G_n(t)$ is a regular function of t and decays as t^{-3} when $t \to \pm\infty$ as shown in Chapter 4. In this case it is convenient to take $t_0 = -\infty$. Then from (4.27) the asymptotic form of x_n for large negative values of time is given by

$$x_n(t) = a_n t + x_{0n} + i/a_n + O(t^{-1}), \qquad t \to -\infty. \tag{3.43}$$

Therefore

$$\operatorname{Im} x_n(-\infty) = 1/a_n > 0, \qquad n = 1, 2, \ldots, N, \tag{3.44}$$

which implies (3.11) by (3.42).

We shall now proceed to the periodic wave case. Instead of (3.9) and (3.10), the appropriate dependent variable transformation is,

$$u(x, t) = \frac{i}{2}\frac{\partial}{\partial x}\ln\frac{f'(x, t)}{f(x, t)}, \tag{3.45}$$

$$f(x, t) \propto \prod_{j=1}^{\infty}[x - x_j(t)], \tag{3.46}$$

$$f'(x, t) \propto \prod_{j=1}^{\infty}[x - x'_j(t)], \tag{3.47}$$

$$\text{Im } x_j(t) > 0, \qquad j = 1, 2, \ldots, \tag{3.48}$$

$$\text{Im } x'_j(t) < 0, \qquad j = 1, 2, \ldots, \tag{3.49}$$

where x_j and x'_j are complex functions of t whose imaginary parts are positive or negative, respectively. Since f and f' are represented by the form of an infinite product, it may be shown that

$$Hu = \frac{i}{2}H\frac{\partial}{\partial x}\ln\frac{f'}{f} = -\frac{1}{2}\frac{\partial}{\partial x}\ln(f'f), \tag{3.50}$$

using (3.12), (3.13), (3.48), and (3.49). Substituting (3.45) and (3.50) into (3.1) yields the bilinearized BO equation

$$iD_t f' \cdot f = D_x^2 f' \cdot f, \tag{3.51}$$

which has the same form as (3.17). Equation (3.51) can be solved by means of a perturbation method, which was used in deriving the N-soliton solution of the KdV equation (see Section 2.2).

The one-periodic wave solution is obtained by taking

$$f' = 1 + \exp(i\xi_1 - \phi_1), \tag{3.52}$$

$$f = 1 + \exp(i\xi_1 + \phi_1), \tag{3.53}$$

with

$$\xi_1 = k_1(x - a_1 t - x_{01}) + \xi_1^{(0)}, \tag{3.54}$$

$$a_1 = k_1 \coth \phi_1, \tag{3.55}$$

where k_1, a_1, and x_{01} are real constants and $\xi_1^{(0)}$ is an arbitrary phase constant. To satisfy conditions (3.48) or (3.49) it is necessary that

$$\phi_1/k_1 > 0. \tag{3.56}$$

Introducing (3.52) and (3.53) into (3.45), we obtain the one-periodic wave solution expressed in the original variable u as

$$u = \frac{(k_1/2)\tanh\phi_1}{1 + \operatorname{sech}\phi_1 \cos\xi_1}. \tag{3.57}$$

This form coincides with the periodic wave solution presented by Benjamin [15] and Ono [17]. The one-soliton solution is derived from (3.57) in the long-wave limit. To show this, keep a_1 and x_{01} finite, choose $\xi_1^{(0)} = \pi$, and take the long-wave limit $k_1 \to 0$. Substituting the expansions

$$\cos\xi_1 = \cos(k_1\theta_1 + \pi) = -1 + \tfrac{1}{2}k_1^2\theta_1^2 + O(k_1^4), \tag{3.58}$$

$$\operatorname{sech}\phi_1 = 1 - \tfrac{1}{2}(k_1/a_1)^2 + O(k_1^4) \tag{3.59}$$

[Eq. (3.59) is a result of (3.55)] into (3.57) we obtain

$$u = a_1/[(a_1\theta_1)^2 + 1], \tag{3.60}$$

in the limit of $k_1 \to 0$, which is the one-soliton solution of the BO equation (3.3).

The N-periodic wave solution of (3.51) may be constructed by the method presented in Section 2.2 and is expressed as [64]

$$f = \sum_{\mu=0,1} \exp\left[\sum_{j=1}^{N} \mu_j(i\xi_j + \phi_j) + \sum_{j<k}^{(N)} \mu_j\mu_k A_{jk}\right], \tag{3.61}$$

$$f' = \sum_{\mu=0,1} \exp\left[\sum_{j=1}^{N} \mu_j(i\xi - \phi_j) + \sum_{j<k}^{(N)} \mu_j\mu_k A_{jk}\right], \tag{3.62}$$

with

$$\xi_j = k_j(x - a_j t - x_{0j}) + \xi_j^{(0)}, \tag{3.63}$$

$$a_j = k_j \coth\phi_j, \tag{3.64}$$

$$\phi_j/k_j > 0, \tag{3.65}$$

$$e^{A_{jl}} = [(a_j - a_l)^2 - (k_j - k_l)^2]/[(a_j - a_l)^2 - (k_j + k_l)^2], \tag{3.66}$$

where $\sum_{\mu=0,1}$ denotes the summation over all possible combinations of $\mu_1 = 0, 1, \mu_2 = 0, 1, \ldots, \mu_N = 0, 1$, and $\sum_{j<k}^{(N)}$ means the summation under the condition $j < k$. The value of u generated from (3.61) and (3.62) is generally a complex quantity. However, a real u is assured by choosing $\xi_j^{(0)}$ as

$$\xi_j^{(0)} = \xi_{j,\,\mathrm{real}}^{(0)} + i\sum_{j\neq k} A_{jk}/2, \qquad j = 1, 2, \ldots, N, \tag{3.67}$$

where $\xi^{(0)}_{j,\,\mathrm{real}}$ ($j = 1, 2, \ldots, N$) are real constants. Indeed, introducing (3.67) into (3.62) yields

$$f' = \exp\left[\sum_{j=1}^{N} (i\xi_j - \phi_j) + \sum_{j<k}^{(N)} A_{jk}\right] f^*, \tag{3.68}$$

which, when substituted into (3.9), gives a real u as

$$u = -\frac{1}{2}\sum_{j=1}^{N} k_j + \frac{i}{2}\frac{\partial}{\partial x}\ln\frac{f^*}{f}. \tag{3.69}$$

To obtain the N-soliton solution in the long-wave limit of the N-periodic wave solution, we set

$$\xi^{(0)}_{j,\,\mathrm{real}} = \pi, \qquad j = 1, 2, \ldots, N, \tag{3.70}$$

in (3.67). From the expansion for small k_j and k_l with finite a_j and a_l,

$$\begin{aligned} A_{jl} &= \ln\frac{(a_j - a_l)^2 - (k_j - k_l)^2}{(a_j - a_l)^2 - (k_j + k_l)^2} \\ &= \frac{4k_j k_l}{(a_j - a_l)^2} + O(k_j^4) = k_j k_l M_{jl}^2 + O(k_j^4), \end{aligned} \tag{3.71}$$

and from (3.61), (3.67), (3.70), and (3.71), in the limit of $k_j, k_l \to 0$, we obtain

$$\begin{aligned} f \simeq \sum_{\mu=0,1} \Bigg\{ &\prod_{j=1}^{N} (-1)^{\mu_j}\left[1 + \mu_j k_j\left(i\theta_j + \frac{1}{a_j}\right)\right] \\ &\times \prod_{j<l}^{(N)}\left[1 + \left(\mu_j\mu_l - \frac{1}{2}\mu_j - \frac{1}{2}\mu_l\right)k_j k_l M_{jl}^2\right]\Bigg\}, \end{aligned} \tag{3.72}$$

where θ_j and M_{jl} are given by (3.8) and (3.38), respectively. When $k_j = 0$ ($j = 1, 2, \ldots, N$)

$$f = \sum_{\mu=0,1} \prod_{j=1}^{N} (-1)^{\mu_j} = 0, \tag{3.73}$$

which means that f is factorized by $\prod_{j=1}^{N} k_j$. Therefore, the leading terms of (3.72) are given by those on the order of $\prod_{j=1}^{N} k_j$ of

$$\prod_{j=1}^{N}\left[1 + k_j\left(i\theta_j + \frac{1}{a_j}\right)\right]\prod_{j<l}^{(N)}[1 + k_j k_l M_{jl}^2], \tag{3.74}$$

yielding the following expression for f in the long-wave limit [64]:

$$f \propto \prod_{j=1}^{N}(i\theta_j + 1/a_j) + \frac{1}{2}\sum_{j,k}^{(N)} M_{jk}^2 \sum_{l \neq j,k}^{N} (i\theta_l + 1/a_l) + \cdots$$

$$+ 1/M!\,2^M \sum_{j,k,\ldots,m,n}^{(N)} M_{jk}^2 \cdots M_{mn}^2 \prod_{p \neq j,k,\ldots,m,n}^{N} (i\theta_p + 1/a_p) + \cdots$$

$$= \begin{vmatrix} i\theta_1 + \dfrac{1}{a_1} & M_{12} & \cdots & M_{1N} \\ M_{21} & i\theta_2 + \dfrac{1}{a_2} & \cdots & M_{2N} \\ \vdots & \vdots & \ddots & \vdots \\ M_{N1} & M_{N2} & \cdots & i\theta_N + \dfrac{1}{a_N} \end{vmatrix}, \tag{3.75}$$

and it follows from (3.69) that

$$u = \frac{i}{2}\frac{\partial}{\partial x}\ln\frac{f^*}{f}. \tag{3.76}$$

in the same limit. Expression (3.76) with (3.75) corresponds to that of the N-soliton solution given by (3.9) together with (3.36)–(3.38).

3.1.2 Derivation of the N-Soliton Solution by the Theory of Linear Algebra

The second method for obtaining the N-soliton solution of the BO equation is from the theory of linear algebra. It is well known that the system of linear algebraic equation is solved by Cramer's formula, which expresses the solution in a determinant form. The systematic use of this formula makes the derivation of the N-soliton solution more transparent and clarifies the structure of the solution.[§]

[§] This discussion follows that of Matsuno [89].

Consider the system of linear algebraic equations for unknowns v_k and $v_k^\dagger$:

$$\sum_{k=1}^{N} M_{jk} v_k = 1, \qquad j = 1, 2, \ldots, N, \tag{3.77}$$

$$\sum_{k=1}^{N} M_{jk}^\dagger v_k^\dagger = -1, \qquad j = 1, 2, \ldots, N, \tag{3.78}$$

where M_{jk} is the (j, k) element of matrix M defined in (3.37) and (3.38) and $M^\dagger$ a matrix with elements

$$M_{jk}^\dagger = M_{kj}^*. \tag{3.79}$$

Matrices M and $M^\dagger$ are nonsingular for real x and t, that is,

$$\det M \neq 0, \tag{3.80}$$

$$\det M^\dagger \neq 0. \tag{3.81}$$

Relation (3.80) is a consequence of (3.10), (3.11), and (3.36), and (3.81) derives from (3.79) and (3.80).

Given relations (3.80) and (3.81), (3.77) and (3.78) are solved uniquely by using Cramer's formula as

$$v_j = \sum_{k=1}^{N} \tilde{M}_{kj}/\det M, \qquad j = 1, 2, \ldots, N, \tag{3.82}$$

$$v_j^\dagger = -\sum_{k=1}^{N} \tilde{M}_{jk}^*/\det M^*, \qquad j = 1, 2, \ldots, N, \tag{3.83}$$

where $\tilde{M}_{jk}$ denotes the cofactor of M_{jk} defined by

$$M\tilde{M} = \tilde{M}M = \mathrm{I}, \qquad I: \text{unit matrix}. \tag{3.84}$$

Then it follows from (3.82) and (3.83) that

$$\sum_{j=1}^{N} v_j^\dagger = -\sum_{j=1}^{N}\sum_{k=1}^{N} \tilde{M}_{jk}^*/\det M^* = -\sum_{j=1}^{N}\left(\sum_{k=1}^{N} \tilde{M}_{jk}/\det M\right)^* = -\sum_{j=1}^{N} v_j^*. \tag{3.85}$$

Similarly, from (3.9) and (3.36)–(3.38),

$$\begin{aligned} u(x, t) &= \frac{1}{2}\left(\sum_{j=1}^{N} \tilde{M}_{jj}/\det M + \sum_{j=1}^{N} \tilde{M}_{jj}^*/\det M^*\right) \\ &= \frac{1}{2}\left(\sum_{j=1}^{N} u_j + \sum_{j=1}^{N} u_j^*\right), \end{aligned} \tag{3.86}$$

with

$$u_j = \tilde{M}_{jj}/\det M, \qquad j = 1, 2, \ldots, N, \tag{3.87}$$

where we have used the property of differentiating a determinant with only the diagonal elements depending on x [see (3.37) and (3.38)]. As a consequence of a remarkable property of the matrix M

$$\sum_{\substack{j,k=1 \\ (j \neq k)}}^{N} \tilde{M}_{jk} = 0, \tag{3.88}$$

which is proved in Appendix II [see (App. II.22)], we obtain the useful relation

$$\sum_{j,k=1}^{N} \tilde{M}_{jk} = \sum_{j=1}^{N} \tilde{M}_{jj} + \sum_{\substack{j,k=1 \\ (j \neq k)}}^{N} \tilde{M}_{jk} = \sum_{j=1}^{N} \tilde{M}_{jj}. \tag{3.89}$$

Then it follows from (3.82), (3.87), and (3.89) that

$$\sum_{j=1}^{N} u_j = \sum_{j=1}^{N} \tilde{M}_{jj}/\det M = \sum_{j=1}^{N} \sum_{k=1}^{N} \tilde{M}_{kj}/\det M = \sum_{j=1}^{N} v_j \tag{3.90}$$

and from (3.85), (3.86), and (3.90) that

$$u(x, t) = \frac{1}{2}\left(\sum_{j=1}^{N} v_j + \sum_{j=1}^{N} v_j^*\right) = \frac{1}{2}\left(\sum_{j=1}^{N} v_j - \sum_{j=1}^{N} v_j^\dagger\right). \tag{3.91}$$

Furthermore, from (3.9) we may express $u(x, t)$ as

$$u(x, t) = \frac{i}{2}\left(\frac{\partial}{\partial x} \ln f^* - \frac{\partial}{\partial x} \ln f\right), \tag{3.92}$$

which, combined with (3.91), yields

$$\sum_{j=1}^{N} v_j = -i \frac{\partial}{\partial x} \ln f, \tag{3.93}$$

$$\sum_{j=1}^{N} v_j^\dagger = -i \frac{\partial}{\partial x} \ln f^*. \tag{3.94}$$

By employing relations

$$H\left(\frac{\partial}{\partial x}\ln f\right) = -i\frac{\partial}{\partial x}\ln f, \tag{3.95}$$

$$H\left(\frac{\partial}{\partial x}\ln f^*\right) = i\frac{\partial}{\partial x}\ln f^*, \tag{3.96}$$

which are derived from (3.10) and (3.12), we obtain from (3.93) and (3.95)

$$H\left(\sum_{j=1}^{N} v_j\right) = -iH\frac{\partial}{\partial x}\ln f = -\frac{\partial}{\partial x}\ln f = -i\sum_{j=1}^{N} v_j, \tag{3.97}$$

and similarly from (3.94) and (3.96)

$$H\left(\sum_{j=1}^{N} v_j^\dagger\right) = i\sum_{j=1}^{N} v_j^\dagger. \tag{3.98}$$

Given these relations, we now prove that $u(x, t)$ given by (3.9) and (3.36)–(3.38) satisfies the BO equation (3.1). For this purpose, consider the quantity

$$A_j \equiv \sum_{k=1}^{N} M_{jk}^\dagger[iv_{k,x}^\dagger - a_k(v_k^\dagger - v_k) + 2uv_k^\dagger], \qquad j = 1, 2, \dots, N. \tag{3.99}$$

Differentiating (3.78) with respect to x gives

$$\sum_{k=1}^{N} M_{jk}^\dagger v_{k,x}^\dagger - iv_j^\dagger = 0, \tag{3.100}$$

due to (3.37) and (3.38). Substituting (3.100) into (3.99), using (3.78) and the explicit expression for $M_{jk}^\dagger$,

$$M_{jk}^\dagger = M_{kj}^* = (-i\theta_j + 1/a_j)\delta_{jk} + [2/(a_k - a_j)](1 - \delta_{jk}), \tag{3.101}$$

where δ_{jk} denotes Kronecker's delta, $\delta_{jk} = 1$ for $j = k$, $\delta_{jk} = 0$ for $j \neq k$, and noting the identity

$$a_k/(a_k - a_j) = 1 + a_j/(a_k - a_j), \tag{3.102}$$

(3.99) can be modified to the form

$$A_j = -v_j^\dagger - \frac{1}{2}\left(-i\theta_j + \frac{1}{a_j}\right)a_j(v_j^\dagger - v_j) - \frac{1}{2}\sum_{\substack{k=1\\(k\neq j)}}^{N} \frac{2a_k}{a_k - a_j}(v_k^\dagger - v_k) - 2u$$

$$= -v_j^\dagger - \frac{1}{2}(v_j^\dagger - v_j) + \frac{i}{2}a_j\theta_j(v_j^\dagger - v_j)$$

$$+ \sum_{\substack{k=1\\(k\neq j)}}^{N} \frac{a_j}{a_j - a_k}(v_k^\dagger - v_k) - \sum_{\substack{k=1\\(k\neq j)}}^{N}(v_k^\dagger - v_k) - 2u. \quad (3.103)$$

To eliminate the third and fourth terms on the right-hand side of (3.103), multiply by a_j on both sides of (3.77) and (3.78) and add the resultant equations to give

$$ia_j\theta_j(v_j - v_j^\dagger) + (v_j + v_j^\dagger) + 2\sum_{\substack{k=1\\(k\neq j)}}^{N} \frac{a_j}{a_j - a_k}(v_k - v_k^\dagger) = 0. \quad (3.104)$$

Substituting (3.104) into (3.103) yields

$$A_j = -v_j^\dagger - \frac{1}{2}(v_j^\dagger - v_j) + \frac{1}{2}(v_j + v_j^\dagger) - \sum_{\substack{k=1\\(k\neq j)}}^{N}(v_k^\dagger - v_k) - 2u$$

$$= -\sum_{k=1}^{N}(v_k^\dagger - v_k) - 2u = 0, \qquad j = 1, 2, \ldots, N, \quad (3.105)$$

where (3.91) has been used in the last step. Relations (3.105) may be identified with the system of homogeneous linear algebraic equations whose unknowns are given by the quantities in parentheses on the right-hand side of (3.99). Hence we conclude from properties of determinants and (3.81) that

$$iv_{j,x}^\dagger - \tfrac{1}{2}a_j(v_j^\dagger - v_j) + 2uv_j^\dagger = 0, \qquad j = 1, 2, \ldots, N. \quad (3.106)$$

Summing (3.106) with respect to j yields

$$i\sum_{j=1}^{N} v_{j,x}^\dagger - \frac{1}{2}\sum_{j=1}^{N} a_j(v_j^\dagger - v_j) + 2u\sum_{j=1}^{N} v_j^\dagger = 0, \quad (3.107)$$

and it follows from the complex conjugate expression of (3.107) together with (3.85) that

$$i\sum_{j=1}^{N} v_{j,x} - \frac{1}{2}\sum_{j=1}^{N} a_j[(v_j^\dagger)^* - v_j^*] - 2u\sum_{j=1}^{N} v_j = 0. \quad (3.108)$$

Adding (3.107) and (3.108) and using (3.91) yields

$$i\sum_{j=1}^{N}(v_{j,x}+v_{j,x}^{\dagger})=\frac{1}{2}\sum_{j=1}^{N}a_j(v_j^{\dagger}-v_j)+\frac{1}{2}\sum_{j=1}^{N}a_j[(v_j^{\dagger})^*-v_j^*]+4u^2. \tag{3.109}$$

Then we have from (3.91), (3.97), (3.98), and (3.109)

$$\begin{aligned}Hu_x&=\frac{1}{2}\left(H\sum_{j=1}^{N}v_{j,x}-H\sum_{j=1}^{N}v_{j,x}^{\dagger}\right)=-\frac{1}{2}i\sum_{j=1}^{N}(v_{j,x}+v_{j,x}^{\dagger})\\&=-\frac{1}{4}\sum_{j=1}^{N}a_j(v_j^{\dagger}-v_j)-\frac{1}{4}\sum_{j=1}^{N}a_j[(v_j^{\dagger})^*-v_j^*]-2u^2,\end{aligned} \tag{3.110}$$

or

$$2u^2+Hu_x=\frac{1}{4}\sum_{j=1}^{N}[a_jv_j-a_j(v_j^{\dagger})^*]+\text{complex conjugate}. \tag{3.111}$$

Substituting (3.82) and (3.83) into (3.111), we obtain

$$\begin{aligned}2u^2+Hu_x&=\sum_{j,k=1}^{N}\frac{\tilde{M}_{kj}a_j+\tilde{M}_{jk}a_j}{4\det M}+\text{complex conjugate}\\&=\sum_{j=1}^{N}\frac{\tilde{M}_{jj}a_j}{2\det M}+\sum_{\substack{j,k=1\\(j\neq k)}}^{N}\frac{\tilde{M}_{kj}a_j+\tilde{M}_{jk}a_j}{4\det M}\\&\quad+\text{complex conjugate}.\end{aligned} \tag{3.112}$$

Since the second term and the complex conjugate expression on the right-hand side of (3.112) vanish identically [see (App. II.23)],

$$\sum_{\substack{j,k=1\\(j\neq k)}}^{N}\frac{\tilde{M}_{kj}a_j+\tilde{M}_{jk}a_j}{4\det M}=0, \tag{3.113}$$

(3.112) reduces to

$$2u^2+Hu_x=\sum_{j=1}^{N}\frac{\tilde{M}_{jj}a_j}{2\det M}+\text{complex conjugate}. \tag{3.114}$$

It follows from (3.36)–(3.38) and properties of determinant differentiation that

$$-\frac{i}{2}\frac{\partial}{\partial t}\ln\frac{f^*}{f}=\sum_{j=1}^{N}\frac{\tilde{M}_{jj}a_j}{2\det M}+\text{complex conjugate}, \tag{3.115}$$

which, compared with (3.114), yields

$$2u^2 + Hu_x = -\frac{i}{2}\frac{\partial}{\partial t}\ln\frac{f^*}{f}. \tag{3.116}$$

Finally, differentiating both sides of (3.116) with respect to x and using the definition of u (3.9), we obtain the equation

$$4uu_x + Hu_{xx} = -u_t, \tag{3.117}$$

which is the BO equation (3.1). Thus, we have completed the proof that expression (3.9) together with (3.36)–(3.38) is an exact N-soliton solution of the BO equation derived from the theory of linear algebra.

Three interesting relations follow from this solution. The first is an expression of $u(x, t)$ in terms of v_j and v_j^* ($j = 1, 2, \ldots, N$), which are solutions of the system of linear algebraic equations (3.77). Multiply v_j on both sides of (3.77), sum with respect to j from 1 to N, and use (3.91) to obtain

$$u(x, t) = \frac{1}{2}\sum_{j=1}^{N}\left(i\theta_j + \frac{1}{a_j}\right)v_j^2 + \text{complex conjugate}, \tag{3.118}$$

where the identity

$$\sum_{\substack{j,k-1 \\ (j\neq k)}}^{N} \frac{v_k v_j}{a_k - a_j} = 0 \tag{3.119}$$

has been used. Expression (3.118) is similar to the KdV N-soliton solution in which u is expressed in terms of the squared eigenfunctions [43].

It follows by differentiating (3.77) s times with respect to x, multiplying by $\partial^s v_j/\partial x^s$, and summing over $j = 1, \ldots, N$ that

$$\sum_{j=1}^{N}\left(i\theta_j + \frac{1}{a_j}\right)\left(\frac{\partial^s v_j}{\partial x^s}\right)^2 + \frac{is}{2}\sum_{j=1}^{N}\frac{\partial}{\partial x}\left(\frac{\partial^{s-1} v_j}{\partial x^{s-1}}\right)^2 = 0, \qquad s = 1, 2, \ldots. \tag{3.120}$$

Integrating (3.120) with respect to x from $-\infty$ to ∞ and using the boundary condition $v_j \to 0$ as $|x| \to \infty$, we obtain the second relation

$$\sum_{j=1}^{N}\int_{-\infty}^{\infty}\left(i\theta_j + \frac{1}{a_j}\right)\left(\frac{\partial^s v_j}{\partial x^s}\right)^2 dx = 0, \qquad s = 1, 2, \ldots. \tag{3.121}$$

The third relation concerns an initial condition which evolves into pure N solitons as time goes to infinity [92]. As shown in (3.91), an N-soliton solution is represented in terms of v_j ($j = 1, 2, \ldots, N$), which are solutions of (3.77). Therefore, it is natural to start with (3.77) to find an initial condition that evolves into pure N solitons. It follows from (3.77) with $t = 0$ that

$$\left[i(x - x_{0j}) + \frac{1}{a_j}\right]v_j(x, 0) + \sum_{\substack{k=1 \\ (k \neq j)}}^{N} \frac{2v_k(x, 0)}{a_j - a_k} = -1. \qquad (3.122)$$

Assume the initial form of $v_j(x, 0)$ to be

$$v_j(x, 0) = (ix + 1)^{-1}, \qquad j = 1, 2, \ldots, N, \qquad (3.123a)$$

and set

$$x_{j0} = 0, \qquad j = 1, 2, \ldots, N, \qquad (3.123b)$$

in (3.122). Substituting (3.123a) and (3.123b) into (3.122) yields the system of N nonlinear equations for the amplitudes a_j:

$$\sum_{\substack{k=1 \\ (k \neq j)}}^{N} \frac{1}{a_j - a_k} = \frac{1}{2}\left(1 - \frac{1}{a_j}\right), \qquad j = 1, 2, \ldots, N. \qquad (3.124)$$

These equations characterize the zeros of the Laguerre polynomial L_N of order N, as shown in Section 3.3 [see (3.297)]. If we identify a_j with the jth zero of L_N, an initial shape of u evolves completely into pure N solitons with the amplitudes equal to the zeros of L_N corresponding to (3.123a) and (3.123b).[§]

Using (3.123a) and (3.91), the initial condition corresponding to (3.123a) and (3.123b) is represented by a simple Lorentzian profile

$$u(x, 0) = \frac{1}{2}\sum_{j=1}^{N}[v_j(x, 0) + v_j^*(x, 0)] = \frac{N}{x^2 + 1}. \qquad (3.125)$$

3.1.3 *Derivation of the N-Soliton Solution by the Pole Expansion Method*

The third method for obtaining the N-soliton solution of the BO equation is the pole expansion method. This method originated in a study of the time evolution of positions of the poles of special solutions

[§] This statement will also be verified by means of a quite different method in Section 3.3.

of the KdV equation; the original concept was described by Kruskal [93]. Thickstun [94] applied it to the KdV soliton solutions and clarified the mechanism of soliton interaction. Airault *el al.* [95] developed the pole expansions of rational and elliptic solutions of the KdV and Boussinesq equations and found a relationship between the motion of the poles of these solutions and the time evolution of certain types of one-dimensional many-body problems. The pole expansion method was then studied extensively and was shown to be applicable to a wide class of nonlinear evolution equations [96, 97].

The pole expansion method was also employed to obtain the N-soliton [90, 91, 98] and the N-periodic wave [99] solutions of the BO equation. We now proceed with a simple derivation of the N-soliton solution of the BO equation as developed by Case [98].

As demonstrated in Section 3.1.1, if the N-soliton solution is assumed to be in form (3.9) together with (3.10), then the time evolution of the nth pole x_n is governed by a nonlinear equation (3.39). Differentiating (3.39) with respect to t and using (3.39) again to eliminate $\dot{x}_s$, $\dot{x}_s^*$, and $\dot{x}_n$,[§] we obtain the second-order equation

$$\frac{d^2 x_n}{dt^2} = 8 \sum_{\substack{s=1 \\ (s \neq n)}}^{N} \frac{1}{(x_s - x_n)^3}$$

$$= -2 \frac{\partial}{\partial x_n} \sum_{\substack{s,j=1 \\ (s \neq j)}}^{N} \frac{1}{(x_s - x_j)^2}, \qquad n = 1, 2, \ldots, N. \quad (3.126)$$

This system of equations is equivalent to an N-body Hamiltonian system interacting with a potential proportional to the inverse square of the distance. It is important to note that (3.126) can be rewritten in the Lax form [100]

$$\partial L/\partial t = BL - LB \equiv [B, L], \quad (3.127)$$

where L and B are the $N \times N$ matrices whose elements are given by

$$L_{lm} = \delta_{lm} \dot{x}_l + (1 - \delta_{lm}) \frac{-2i}{x_l - x_m}, \qquad \dot{x}_l \equiv \frac{dx_l}{dt}, \quad (3.128)$$

$$B_{lm} = \delta_{lm} \sum_{\substack{s=1 \\ (s \neq l)}}^{N} \frac{2i}{(x_l - x_s)^2} + (1 - \delta_{lm}) \frac{-2i}{(x_l - x_m)^2}, \quad (3.129)$$

which may be easily verified by direct calculation.

[§] $\dot{x}_s = dx_s/dt$.

If we now define the $N \times N$ matrix $K[x(t)]$ with elements

$$K_{jl} = \delta_{jl} x_l(t), \tag{3.130}$$

then it follows that [98]

$$K[x(t)] = U\{K[x(t_0)] + (t - t_0)L[x(t_0), \dot{x}(t_0)]\}U^{-1}, \tag{3.131}$$

where the $N \times N$ matrix U is given by a solution of the evolution equation

$$\partial U/\partial t = BU, \qquad U(t_0) = I, \quad I: \text{unit matrix}, \tag{3.132}$$

and $x(t_0)$ and $\dot{x}(t_0)$ are vectors with components

$$x(t_0) = [x_1(t_0), x_2(t_0), \ldots, x_N(t_0)], \tag{3.133}$$

$$\dot{x}(t_0) = [\dot{x}_1(t_0), \dot{x}_2(t_0), \ldots, \dot{x}_N(t_0)], \tag{3.134}$$

where the conditions imposed on x_j are

$$\operatorname{Im} x_j(t_0) > 0, \qquad j = 1, 2, \ldots, N \tag{3.135}$$

$$x_j(t_0) \neq x_l(t_0) \qquad \text{for} \quad j \neq l. \tag{3.136}$$

To prove (3.131), consider the quantity

$$J(t) = U^{-1}K[x(t)]U. \tag{3.137}$$

Differentiating (3.137) with respect to t gives

$$\partial J/\partial t = U^{-1}\{K[\dot{x}(t)] + [K, B]\}U \tag{3.138}$$

by (3.132). Using the relation [95],

$$[K, B] = L - K[\dot{x}(t)], \tag{3.139}$$

(3.138) reduces to

$$\partial J/\partial t = U^{-1}LU. \tag{3.140}$$

Differentiating (3.140) once more with respect to t and using (3.127), we find

$$\partial^2 J/\partial t^2 = U^{-1}\{\partial L/\partial t - [B, L]\}U = 0. \tag{3.141}$$

Therefore, J has a form

$$J(t) = c_1 + c_2(t - t_0), \tag{3.142}$$

where c_1 and c_2 are constant matrices. Setting $t = t_0$ in (3.137),

$$c_1 = K[x(t_0)] \tag{3.143}$$

by (3.132). Matrix c_2 is also determined from (3.140) with $t = t_0$ as

$$c_2 = L[x(t_0), \dot{x}(t_0)]. \qquad (3.144)$$

Thus, $J(t)$ becomes

$$J(t) = K[x(t_0)] + (t - t_0)L[x(t_0), \dot{x}(t_0)], \qquad (3.145)$$

which, combined with (3.137), yields relation (3.131). The N-soliton solution is now constructed as

$$\begin{aligned}
\prod_{j=1}^{N}(x - x_j) & \\
&= \det\{xI - K[x(t)]\}, \quad \text{by the definition of } K[x(t)], \\
&= \det U^{-1}\{xI - K[x(t)]\}U, \quad \text{by } \det AB = \det BA, \\
&= \det\{xI - U^{-1}K[x(t)]U\}, \quad \text{by } U^{-1}U = I, \\
&= \det\{xI - K[x(t_0)] - (t - t_0)L[x(t_0), \dot{x}(t_0)]\}, \quad \text{by (3.131)}, \\
&= \det\{xI - Z\}, \qquad (3.146)
\end{aligned}$$

where Z is the $N \times N$ matrix with elements

$$Z_{lm} = \delta_{lm}x_l(t_0) + (t - t_0)L_{lm}(t_0). \qquad (3.147)$$

It follows from (3.9), (3.10), and (3.146) that

$$\begin{aligned}
u &= \frac{i}{2}\frac{\partial}{\partial x}\ln\left[\prod_{j=1}^{N}(x - x_j^*) \Big/ \prod_{j=1}^{N}(x - x_j)\right] \\
&= \frac{i}{2}\frac{\partial}{\partial x}\ln[\det(xI - Z^*)/\det(xI - Z)], \qquad (3.148)
\end{aligned}$$

which is an explicit expression of the N-soliton solution.

Expression (3.148) may be reduced to that of the N-soliton solution derived by the bilinear transformation method. Assume an asymptotic form of $x_j(t)$ for large t as

$$x_j(t) \simeq a_j t + i/a_j + x_{0j}, \qquad t \to \infty, \quad j = 1, 2, \ldots, N. \qquad (3.149)$$

Taking the limit $t_0 \to \infty$, (3.147) becomes

$$\begin{aligned}
Z_{lm} &= \delta_{lm}(a_j t_0 + i/a_j + x_{0j}) + \delta_{lm}(t - t_0)a_j + (1 - \delta_{lm})2i/(a_l - a_m) \\
&= \delta_{lm}(a_j t + i/a_j + x_{0j}) + (1 - \delta_{lm})2i/(a_l - a_m). \qquad (3.150)
\end{aligned}$$

Substituting (3.150) into (3.148), we obtain (3.9) with (3.36)–(3.38).

The pole expansion method can also be applied to obtain the N-soliton solutions of the higher-order BO equations [101] and the N-periodic wave solution of the BO equation [99, 102].

3.2 Bäcklund Transformation and Conservation Laws of the Benjamin–Ono Equation

In this section the Bäcklund transformation of the BO equation is formulated on the basis of the bilinear transformation method. An infinite number of conservation laws of the BO equation are then constructed from the Bäcklund transformation of the BO equation and the structure of conserved quantity is clarified.

3.2.1 Bäcklund Transformation

We first write the BO equation in the convenient form

$$u_t + 4uu_x + \beta H u_{xx} = 0, \tag{3.151}$$

where β (>0) is a parameter characterizing the magnitude of the dispersion. Introducing the dependent variable transformation

$$u = \frac{i}{2}\beta\frac{\partial}{\partial x}\ln\frac{f'}{f}, \tag{3.152}$$

where f is given by (3.46) and f' by (3.47), (3.151) is transformed into the bilinear equation

$$iD_t f'\cdot f = \beta D_x^2 f'\cdot f. \tag{3.153}$$

Let v be another solution of (3.151), that is,

$$v = \frac{i}{2}\beta\frac{\partial}{\partial x}\ln\frac{g'}{g}, \tag{3.154}$$

$$iD_t g'\cdot g = \beta D_x^2 g'\cdot g. \tag{3.155}$$

The relation connecting the two solutions u and v, the Bäcklund transformation of the BO equation, is given in terms of the bilinear variables as [103]

$$(iD_t - 2i\lambda D_x - \beta D_x^2 - \mu)f \cdot g = 0, \tag{3.156}$$

$$(iD_t - 2i\lambda D_x - \beta D_x^2 - \mu)f' \cdot g' = 0, \tag{3.157}$$

$$(\beta D_x + i\lambda)f \cdot g' = i\nu f'g, \tag{3.158}$$

where λ, μ, and ν are arbitrary constants. If f and f' satisfy (3.153) then g and g' also satisfy (3.155), provided that Eqs. (3.156)–(3.158) hold for f, f', g, and g'. We show this by considering the quantity

$$P \equiv [(iD_t - \beta D_x^2)f' \cdot f]g'g - f'f[(iD_t - \beta D_x^2)g' \cdot g]. \tag{3.159}$$

Using formula (App. I.6.1), P is converted to

$$P = (iD_t f' \cdot g')fg - f'g'(iD_t f \cdot g) - \beta(D_x^2 f' \cdot f)g'g + \beta f'f(D_x^2 g' \cdot g). \tag{3.160}$$

Substituting (3.156) and (3.157) into (3.160) and using formulas (App. I.6.1) and (App. I.6.3), P becomes

$$\begin{aligned} P &= 2i\lambda[(D_x f' \cdot g')fg - f'g'(D_x f \cdot g)] \\ &\quad + \beta[(D_x^2 f' \cdot g')fg - f'g'(D_x^2 f \cdot g)] \\ &\quad - \beta[(D_x^2 f' \cdot f)g'g - f'f(D_x^2 g' \cdot g)] \\ &= 2i\lambda D_x f'g \cdot g'f + \beta D_x[D_x(f' \cdot g) \cdot fg' + f'g \cdot (D_x f \cdot g')] \\ &\quad - \beta D_x[D_x(f' \cdot g) \cdot g'f + f'g \cdot (D_x g' \cdot f)] \\ &= 2D_x[f'g \cdot (i\lambda + \beta D_x)f \cdot g']. \end{aligned} \tag{3.161}$$

This last expression vanishes identically owing to (3.158) and (App. I.1.5), that is,

$$P = 0. \tag{3.162}$$

Therefore, (3.155) follows from (3.162) and (3.153), and the proof is complete. The superposition formula and the commutability relation may be derived similarly on the basis of (3.156)–(3.158), which has been detailed by Nakamura [103].

To transform (3.156)–(3.158) into a form written in the original variables, we introduce the potential functions $\bar{u}$ and $\bar{v}$ through the relations

$$u = \bar{u}_x, \tag{3.163}$$

$$v = \bar{v}_x, \tag{3.164}$$

or

$$\bar{u} = \frac{i\beta}{2} \ln \frac{f'}{f}, \tag{3.165}$$

$$\bar{v} = \frac{i\beta}{2} \ln \frac{g'}{g} \tag{3.166}$$

from (3.152) and (3.154), respectively. Note also the relations that are derived using (3.50):

$$\frac{1}{2}(1 + iH)u = \frac{1}{2}\left[\frac{i\beta}{2}\left(\frac{f'_x}{f'} - \frac{f_x}{f}\right) - \frac{i\beta}{2}\left(\frac{f'_x}{f'} + \frac{f_x}{f}\right)\right] = -\frac{i\beta}{2}\frac{f_x}{f}, \tag{3.167}$$

and

$$\frac{1}{2}(1 - iH)u = \frac{i\beta}{2}\frac{f'_x}{f'}. \tag{3.168}$$

Dividing (3.158) by fg' yields

$$\beta\left(\frac{f_x}{f} - \frac{g'_x}{g'}\right) + i\lambda = i\nu \frac{f'g}{fg'}. \tag{3.169}$$

Substituting (3.163)–(3.168) into (3.169), we obtain the space part of the Bäcklund transformation written in original variables as

$$(\bar{u} + \bar{v})_x = -\lambda + \nu \exp[-2i(\bar{u} - \bar{v})/\beta] - iH(\bar{u} - \bar{v})_x. \tag{3.170}$$

Introducing a function w by

$$\bar{u} - \bar{v} = -i\beta w/2 \tag{3.171}$$

and ε by

$$\lambda = \nu = -2/\varepsilon, \tag{3.172}$$

(3.170) becomes[§]

$$-(i\beta/2)P_- w_x + (1 - e^{-w})/\varepsilon = u, \tag{3.173}$$

[§] Equation (3.173) will be used extensively in this text.

where P_- is an operator defined by

$$P_- = \tfrac{1}{2}(1 - iH). \tag{3.174}$$

The time part of the Bäcklund transformation, (3.156) and (3.157), is rewritten as

$$\left(i\frac{\partial}{\partial t} - 2i\lambda\frac{\partial}{\partial x}\right)\ln\frac{f}{g} - \beta\frac{\partial^2}{\partial x^2}\ln fg - \beta\left(\frac{\partial}{\partial x}\ln\frac{f}{g}\right)^2 - \mu = 0, \tag{3.175}$$

$$\left(i\frac{\partial}{\partial t} - 2i\lambda\frac{\partial}{\partial x}\right)\ln\frac{f'}{g'} - \beta\frac{\partial^2}{\partial x^2}\ln f'g' - \beta\left(\frac{\partial}{\partial x}\ln\frac{f'}{g'}\right)^2 - \mu = 0. \tag{3.176}$$

Subtracting (3.175) from (3.176) yields

$$\begin{aligned}\left(i\frac{\partial}{\partial t} - 2i\lambda\frac{\partial}{\partial x}\right)\left(\ln\frac{f'}{f} - \ln\frac{g'}{g}\right) - \beta\frac{\partial^2}{\partial x^2}\left(\ln\frac{f'}{f} + \ln\frac{g'}{g}\right)\\ -\beta\left(\frac{\partial}{\partial x}\ln\frac{f'}{f} - \frac{\partial}{\partial x}\ln\frac{g'}{g}\right)\left(\frac{\partial}{\partial x}\ln f'f - \frac{\partial}{\partial x}\ln g'g\right) = 0.\end{aligned} \tag{3.177}$$

Introducing (3.165)–(3.168) into (3.177), we obtain the time part of the Bäcklund transformation written in original variables as

$$(\bar{u} - \bar{v})_t = 2\lambda(\bar{u} - \bar{v})_x + 2i(\bar{u} - \bar{v})_x H(\bar{u} - \bar{v})_x - i\beta(\bar{u} + \bar{v})_{xx}. \tag{3.178}$$

Eliminating the term $(\bar{u} + \bar{v})_{xx}$ by using (3.170) gives

$$\begin{aligned}(\bar{u} - \bar{v})_t = {}& 2\lambda(u - v)_x + 2i(\bar{u} - \bar{v})_x H(\bar{u} - \bar{v})_x\\ & - 2\nu(\bar{u} - \bar{v})_x \exp[-2i(u - v)/\beta] - \beta H(\bar{u} - \bar{v})_{xx}.\end{aligned} \tag{3.179}$$

Finally, by introducing (3.171) and (3.172) into (3.179) it follows that

$$w_t = -\beta H w_{xx} - \frac{4}{\varepsilon}(1 - e^{-w})w_x + \beta w_x H w_x. \tag{3.180}$$

It is interesting to note the relation

$$\begin{aligned}u_t + 4uu_x + \beta H u_{xx} = {}& \left[-\frac{i\beta}{2}P_-\frac{\partial}{\partial x} + \frac{1}{\varepsilon}e^{-w}\right]\\ & \times\left[w_t + \beta H w_{xx} + \frac{4}{\varepsilon}(1 - e^{-w})w_x - \beta w_x H w_x\right],\end{aligned} \tag{3.181}$$

which follows from u in (3.173) and the properties of the H operator (see Appendix III). Therefore, (3.173) and (3.180) imply (3.151). If we expand w formally in powers of ε as

$$w = \sum_{n=1}^{\infty} w_n \varepsilon^n, \tag{3.182}$$

then it can be seen that (3.151) and (3.181) also imply (3.180). The forms (3.173) and (3.180) correspond to those of the Bäcklund transformation of the KdV equation, (2.230) and (2.233).

3.2.2 *Conservation Laws*

It follows by integrating (3.180) from $-\infty$ to ∞ and using a property of the H operator, (App. III.18), together with the boundary condition $w \to 0$ as $|x| \to \infty$ that

$$\frac{d}{dt}\int_{-\infty}^{\infty} w\,dx = 0, \tag{3.183}$$

which means that the function w is a conserved density. Substituting (3.182) into (3.183), we obtain

$$\sum_{n=1}^{\infty} \frac{dI_n}{dt}\varepsilon^n = 0, \tag{3.184}$$

where

$$I_n = \int_{-\infty}^{\infty} w_n\,dx. \tag{3.185}$$

Relation (3.184) must hold for arbitrary ε, therefore

$$dI_n/dt = 0, \qquad n = 1, 2, \ldots . \tag{3.186}$$

The I_n defined in (3.185) is the nth conserved quantity of the BO equation. To derive the explicit functional form of w_n, we introduce (3.182) into (3.173) and compare the ε^n term on both sides of (3.173).

The first few expressions of I_n constructed from these w_n are given as

$$I_1 = \int_{-\infty}^{\infty} u\,dx, \tag{3.187}$$

$$I_2 = \int_{-\infty}^{\infty} \frac{u^2}{2}\,dx, \tag{3.188}$$

$$I_3 = \int_{-\infty}^{\infty} \left(\frac{u^3}{3} + \frac{\beta}{4} uHu_x\right) dx, \tag{3.189}$$

$$I_4 = \int_{-\infty}^{\infty} \left[\frac{u^4}{4} + \frac{3\beta}{8} u^2Hu_x + \frac{\beta^2}{8}(u_x)^2\right] dx, \tag{3.190}$$

$$\vdots$$

$$I_n = \int_{-\infty}^{\infty} \left[\frac{u^n}{n} + \frac{\beta}{4}\sum_{m=1}^{n-2} u^{n-m-1}H(u_x u^{m-1}) + O(\beta^2)\right] dx, \qquad n \geq 3. \tag{3.191}$$

As seen from (3.173) with $\beta = 0$,

$$w = -\ln(1 - \varepsilon u) = \sum_{n=1}^{\infty} \frac{u^n}{n}\varepsilon^n, \tag{3.192}$$

implying that the term that does not contain β has the form u^n/n in the expression of w_n.

It may be informative to discuss another construction of the Bäcklund transformation described by Bock and Kruskal [104]. They considered the *associated linear equation*

$$q_t + 4uq_x + \beta Hq_{xx} = 0 \tag{3.193}$$

and observed that the conserved density for the BO equation

$$q^{(0)} = u, \tag{3.194}$$

$$q^{(1)} = u^2 + \tfrac{1}{2}\beta Hu_x, \tag{3.195}$$

$$q^{(2)} = u^3 + \beta(\tfrac{3}{4}uHu_x + \tfrac{3}{4}Huu_x) - \tfrac{1}{4}\beta^2 u_{xx}, \tag{3.196}$$

$$\vdots$$

is satisfied by (3.193). They then assumed the existence of an infinite series

$$\varepsilon q = \varepsilon \sum_{n=0}^{\infty} q_n \varepsilon^n$$
$$= \varepsilon b_1 u + \varepsilon^2 b_2(u^2 + \tfrac{1}{2}\beta H u_x) + \varepsilon^3 b_3[u^3 + \beta(\tfrac{3}{4}uHu_x + \tfrac{3}{4}Huu_x)$$
$$- \tfrac{1}{4}\beta^2 u_{xx}] + \cdots, \tag{3.197}$$

where b_j $(j = 1, 2, \ldots)$ are unknown constants. Solving (3.197) inversely for u under the condition that this inverse contains no derivative of q higher than the first, they found an expression

$$u = (q - \tfrac{1}{4}\varepsilon\beta H q_x)/(1 + \varepsilon q) - H[\tfrac{1}{4}\varepsilon\beta q_x/(1 + \varepsilon q)], \tag{3.198}$$

which is the analogue of the Miura transformation [105] of the KdV equation. By substituting (3.198) into (3.193), it is confirmed that q is a conserved density, that is,

$$\frac{d}{dt}\int_{-\infty}^{\infty} q\, dx = 0. \tag{3.199}$$

Introducing expansion (3.197) into (3.198) and comparing the ε^n term on both sides of (3.198), the q_n $(n = 1, 2, \ldots)$ are determined successively, the first few corresponding to (3.194)–(3.196).

We now reconsider the Bäcklund transformations (3.173) and (3.180). All information concerning the conservation laws is included in (3.173) and therefore the study of (3.173) may help to clarify their structure. However, (3.173) is highly nonlinear, owing to the term e^{-w}, and hence intractable in the present form. To overcome this, we differentiate (3.173) with respect to x to obtain

$$-\frac{i\beta}{2} P_- w_{xx} + \frac{1}{\varepsilon} e^{-w} w_x = u_x. \tag{3.200}$$

Eliminating the term e^{-w} by (3.173) yields

$$w_x = \varepsilon\left(u_x + u w_x + \frac{i\beta}{2} P_- w_{xx} + \frac{i\beta}{2} w_x P_- w_x\right). \tag{3.201}$$

Introducing (3.182) into (3.201) and comparing the ε^n term on both sides of (3.201), we obtain a recurrence formula for w_n as [106]

$$w_1 = u, \tag{3.202}$$

$$w_{n+1,x} = uw_{n,x} + \frac{i\beta}{2} P_- w_{n,xx} + \frac{i\beta}{2} \sum_{s=1}^{n-1} w_{s,x} P_- w_{n-s,x}, \qquad n \geq 1. \tag{3.203}$$

This formula is a starting point for the following discussion. The first few expressions of w_n derived from (3.202) and (3.203) are

$$w_2 = \frac{u^2}{2} + \frac{i\beta}{2} P_- u_x, \tag{3.204}$$

$$w_3 = \frac{u^3}{3} + \frac{i\beta}{2}(uP_- u_x + P_- uu_x) - \frac{\beta^2}{4} P_- u_{xx}, \tag{3.205}$$

$$\begin{aligned} w_4 = {} & \frac{u^4}{4} + \frac{i\beta}{2}(P_- u^2 u_x + uP_- uu_x + u^2 P_- u_x) \\ & - \frac{\beta^2}{4}\left[P_-(uP_- u_x)_x + P_-(uu_x)_x + uP_- u_{xx} + \frac{1}{2}(P_- u_x)^2\right] \\ & - \frac{i\beta^3}{8} P_- u_{xxx}, \end{aligned} \tag{3.206}$$

which reduce to the conserved quantities (3.188)–(3.190) after integrating with respect to x. Since (3.203) includes a derivative term $w_{n+1,x}$, it is not clear whether (3.203) is integrable. To see this, it is convenient to introduce C_j $(j \geq 1)$ by

$$C_1 = w_1, \tag{3.207}$$

$$C_j = w_j - \frac{i\beta}{2} P_- w_{j-1,x}, \qquad j \geq 2. \tag{3.208}$$

Equation (3.203) is then rewritten in terms of C_j as

$$C_{n+1,x} = \sum_{s=1}^{n} w_{s,x} w_{n-s+1} - \sum_{s=1}^{n-1} w_{s,x} C_{n-s+1}, \qquad n \geq 1, \tag{3.209}$$

which is linear in C_j ($j = 1, 2, \ldots, n + 1$), and its solution is obtained explicitly as

$$C_{n+1} = \sum_{j=1}^{n} \frac{(-1)^{j+1}}{(j+1)!} \sum_{s_1=1}^{n-j+1} \sum_{s_2=1}^{n-j+2-s_1} \sum_{s_3=1}^{n-j+3-s_1-s_2} \cdots$$
$$\times \sum_{s_j=1}^{n-s_1-s_2-\cdots-s_{j-1}} w_{s_1} w_{s_2} \cdots w_{s_j} w_{n-s_1-s_2-\cdots-s_j+1}, \qquad n \geq 1. \tag{3.210}$$

Terms w_{n+1} are constructed recursively from (3.207), (3.208), and (3.210). From (3.207) and (3.208) it may be noted that

$$I_n = \int_{-\infty}^{\infty} w_n \, dx = \int_{-\infty}^{\infty} C_n \, dx, \qquad n = 1, 2, \ldots. \tag{3.211}$$

The important relations which hold between conserved quantities follow from the recurrence formula (3.203). Some notation must be introduced before deriving these relations. The functional derivative $\delta I_n/\delta u$ is defined in the relation

$$\frac{\partial}{\partial \varepsilon} I_n[u + \varepsilon v]|_{\varepsilon=0} = \int_{-\infty}^{\infty} \frac{\delta I_n}{\delta u} v \, dx. \tag{3.212}$$

The derivative D/Du is

$$\frac{D}{Du} = \sum_{n=0}^{\infty} (-1)^n \frac{\partial^n}{\partial x^n} \frac{\partial}{\partial u_{nx}}, \qquad u_{nx} = \frac{\partial^n u}{\partial x^n}, \tag{3.213}$$

where the action of the operator $\partial/\partial u_{nx}$ on Hf, f being an arbitrary function of $u, u_x, \ldots$, is defined by

$$\partial Hf/\partial u_{nx} = H \, \partial f/\partial u_{nx}, \qquad n = 0, 1, 2, \ldots. \tag{3.214}$$

By using a property of the H operator [see (App. III.4)]

$$f Hg = -gHf + H[fg - (Hf)(Hg)], \tag{3.215}$$

we may derive a relation

$$\delta I_n/\delta u = Dw_n/Du + H\tilde{w}_n, \tag{3.216}$$

where $\tilde{w}$ is a certain function of $u, u_x, \ldots, Hu, Hu_x, \ldots$.

We now prove the first relation

$$\partial w_j/\partial u = (j - 1)w_{j-1}, \qquad j \geq 2, \tag{3.217}$$

which is verified by mathematical induction. If we assume (3.217) up to $j = n$, then it follows by differentiating (3.203) with respect to u that

$$\begin{aligned}\frac{\partial}{\partial x}\frac{\partial w_{n+1}}{\partial u} &= w_{n,x} + u\frac{\partial}{\partial x}\frac{\partial w_n}{\partial u} + \frac{i\beta}{2}P_-\frac{\partial^2}{\partial x^2}\frac{\partial w_n}{\partial u} \\ &\quad + \frac{i\beta}{2}\sum_{s=1}^{n-1}\left[\left(\frac{\partial}{\partial x}\frac{\partial w_s}{\partial u}\right)P_- w_{n-s,x} + w_{s,x}P_-\frac{\partial}{\partial x}\left(\frac{\partial w_{n-s}}{\partial u}\right)\right] \\ &= w_{n,x} + (n-1)uw_{n-1,x} + (n-1)\frac{i\beta}{2}P_- w_{n-1,xx} \\ &\quad + \frac{i\beta}{2}\sum_{s=1}^{n-1}\Big[(s-1)w_{s-1,x}P_- w_{n-s,x} \\ &\quad + (n-s-1)w_{s,x}P_- w_{n-s-1,x}\Big],\end{aligned} \tag{3.218}$$

where we have used a commutation relation

$$\frac{\partial}{\partial u}\frac{\partial}{\partial x}f = \frac{\partial}{\partial x}\frac{\partial}{\partial u}f, \tag{3.219}$$

where f is an arbitrary function of $u, u_x, \ldots,$ and (3.217) is used in passing to the second equality of (3.218). Substituting the relations

$$\sum_{s=1}^{n-1}(s-1)w_{s-1,x}P_- w_{n-s,x} = \sum_{s=1}^{n-2} s w_{s,x}P_- w_{n-s-1,x}, \tag{3.220}$$

$$\sum_{s=1}^{n-1}(n-s-1)w_{s,x}P_- w_{n-s-1,x} = \sum_{s=1}^{n-2}(n-s-1)w_{s,x}P_- w_{n-s-1} \tag{3.221}$$

into (3.218), we obtain

$$\begin{aligned}\frac{\partial}{\partial x}\frac{\partial w_{n+1}}{\partial x} &= w_{n,x} + (n-1)\Bigg(uw_{n-1,x} + \frac{i\beta}{2}P_- w_{n-1,xx} \\ &\quad + \frac{i\beta}{2}\sum_{s=1}^{n-2} w_{s,x}P_- w_{n-s-1,x}\Bigg) \\ &= w_{n,x} + (n-1)w_{n,x} = nw_{n,x}, \qquad \text{by (3.203).}\end{aligned} \tag{3.222}$$

Integrating (3.222) with respect to x and using the boundary condition $w_n \to 0$ as $|x| \to \infty$, we have

$$\partial w_{n+1}/\partial u = n w_n, \tag{3.223}$$

which implies that relation (3.217) holds for $j = n + 1$, completing the proof.

The second relation to be proved is

$$\int_{-\infty}^{\infty} \frac{\delta I_n}{\delta u}\,dx = (n-1)I_{n-1}, \qquad n \geq 2, \tag{3.224}$$

which is a consequence of the important fact that the functional derivative of a conserved quantity is a conserved density. To verify this statement, we use (3.213), (3.216), and (3.217) to obtain

$$\begin{aligned}\frac{\delta I_n}{\delta u} &= \frac{\partial w_n}{\partial u} + \sum_{s=1}^{\infty}(-1)^s \frac{\partial^s}{\partial x^s}\frac{\partial w_n}{\partial u_{sx}} + H\tilde{w}_n \\ &= (n-1)w_{n-1} + \sum_{s=1}^{\infty}(-1)^s \frac{\partial^s}{\partial x^s}\frac{\partial w_n}{\partial u_{sx}} + H\tilde{w}_n.\end{aligned} \tag{3.225}$$

Integrating (3.225) with respect to x yields (3.224), by noting the definition of I_n, (3.185), and formula (App. III.2),

$$\int_{-\infty}^{\infty} H\tilde{w}_n\,dx = 0. \tag{3.226}$$

The third relation is

$$K[u+\lambda] = \sum_{s=0}^{n-1} \frac{(n-1)!}{s!(n-s-1)!} K_{n-s}[u]\lambda^s, \qquad n = 1, 2, \ldots, \tag{3.227}$$

where

$$K_n[u] = \delta I_n/\delta u \tag{3.228}$$

and λ is an arbitrary constant. This relation follows by repeated use of (3.224).[§]

Finally, we shall comment on the inverse scattering transform of the BO equation. The derivation of the inverse scattering transform can be

[§] Recent topics concerning the conservation laws of the BO equation appear in the references [107–109]. In [107], an infinite number of conservation laws were constructed from Lie algebra. The conserved quantities which depend explicitly on time were presented in [108, 109] as an application of this theory.

performed formally from the Bäcklund transformation once the Bäcklund transformation has been constructed. (The procedure is the same as that for the KdV equation demonstrated in Section 2.4.) We define wave functions ψ and ψ' by

$$\psi = g/f, \tag{3.229}$$

$$\psi' = g'/f' \tag{3.230}$$

and introduce these into (3.175), (3.176), and (3.169). Taking relations (3.167) and (3.168) into account, we obtain the inverse scattering transform of the BO equation in the form [103]

$$(i\,\partial/\partial t - 2i\lambda\,\partial/\partial x + \beta\,\partial^2/\partial x^2 + 2iu_x - 2Hu_x + \mu)\psi = 0, \tag{3.231}$$

$$(i\,\partial/\partial t - 2i\lambda\,\partial/\partial x + \beta\,\partial^2/\partial x^2 - 2iu_x - 2Hu_x + \mu)\psi' = 0, \tag{3.232}$$

$$(i\beta\,\partial/\partial x + 2u + \lambda)\psi' - \nu\psi = 0. \tag{3.233}$$

One can verify that eliminating ψ and ψ' from (3.231)–(3.233) reproduces the original *BO* equation (3.151). At the present time, however, the standard technique of the inverse scattering method [3, 4] cannot be applied directly to the system of equations (2.231)–(2.233), and we shall not pursue this problem further.

3.3 Asymptotic Solutions of the Benjamin–Ono Equation

The methods developed in the preceding sections are convenient tools for obtaining special solutions of the BO equation. The N-soliton and N-periodic wave solutions are very important and are characteristic of integrable nonlinear evolution equations. From the mathematical point of view, however, the initial value problem must be studied for full understanding of the general nature of solutions.

In this section we shall develop an approximate method for solving an initial value problem of the BO equation and investigate the asymptotic behavior of solutions in the zero dispersion limit. An explicit example of an initial condition which evolves into pure N solitons is also presented, and it is shown that the amplitudes of solitons are then closely related to the zeros of the Laguerre polynomial of order N.[§]

[§] The method presented in this section was developed by Matsuno [110–112].

3.3.1 Method of Exact Solution

We now consider the initial condition

$$u(x, 0) = u_0 \phi(\xi), \qquad \xi = x/l, \tag{3.234}$$

$$\phi(\xi) > 0, \qquad \phi(\pm\infty) = 0, \tag{3.235}$$

where u_0 and l are representative values of $u(x, 0)$ and x, respectively, and ϕ is a dimensionless function characterizing the profile of the initial disturbance. Tentatively assume that the initial condition given by (3.234) and (3.235) evolves into pure N solitons for large values of time. The amplitude of each soliton may be determined from the system of equations derived from an infinite number of conservation laws of the BO equation. To show this, consider first the two-soliton case. For large t, the solution is approximated by

$$u(x, t) \simeq \sum_{j=1}^{2} \frac{a_j}{(a_j/\beta)^2(x - a_j t - x_{0j})^2 + 1}, \tag{3.236}$$

a superposition of two solitons with amplitudes a_1 and a_2. To determine a_1 and a_2, estimate the conserved quantities I_2 and I_3 given by (3.188) and (3.189) at both $t = 0$ and $t \to \infty$. Keeping (3.236) in mind, we obtain

$$\frac{\pi\beta}{4}(a_1 + a_2) = I_2[u(x, 0)] = \int_{-\infty}^{\infty} w_2(x, 0)\, dx, \tag{3.237}$$

$$\frac{\pi\beta}{16}(a_1^2 + a_2^2) = I_3[u(x, 0)] = \int_{-\infty}^{\infty} w_3(x, 0)\, dx. \tag{3.238}$$

Introducing the notation

$$y_j = a_j/4u_0, \qquad j = 1, 2, \ldots, \tag{3.239}$$

$$\tilde{w}_n(\xi, 0) = w_n(u)|_{u=\phi(\xi),\, \beta=\sigma^{-1}}, \qquad n = 1, 2, \ldots, \tag{3.240}$$

$$s_n = \frac{\sigma}{\pi}\int_{-\infty}^{\infty} \tilde{w}_n(\xi, 0)\, d\xi, \qquad n = 1, 2, \ldots, \tag{3.241}$$

$$\sigma = u_0 l/\beta, \tag{3.242}$$

(3.237) and (3.238) can be rewritten as

$$y_1 + y_2 = s_1, \tag{3.243}$$

$$y_1^2 + y_2^2 = s_2. \tag{3.244}$$

Values y_1 and y_2 are determined from the roots of an algebraic equation of order two

$$y^2 - s_1 y + \tfrac{1}{2}(s_1^2 - s_2) = 0 \tag{3.245}$$

as

$$y_1 = \tfrac{1}{2}(s_1 - \sqrt{2s_2 - s_1^2}), \tag{3.246}$$

$$y_2 = \tfrac{1}{2}(s_1 + \sqrt{2s_2 - s_1^2}), \tag{3.247}$$

where we have assumed $y_1 < y_2$. To yield real and positive y_1 and y_2, it is necessary that the following conditions are satisfied:

$$s_1^2 > s_2, \tag{3.248}$$

$$s_2 > s_1^2/2. \tag{3.249}$$

This discussion may be extended to the general N-soliton case. In this case, however, we must evaluate the nth conserved quantity I_n for pure N-soliton solution. For this purpose, we employ the Bäcklund transformation of the BO equation (3.173) and (3.180).

We first evaluate I_n for the one-soliton solution (3.3). Note that the relation

$$w_t[u_s(x, t)] = -aw_x[u_s(x, t)] \tag{3.250}$$

holds since $u_s(x, t)$ is a function of $x - at$. Substituting (3.250) into (3.180), integrating from $-\infty$ to x, and using the boundary condition $w \to 0$ as $|x| \to \infty$, we obtain

$$\left(\frac{4}{\varepsilon} - a\right)w + \beta H w_x - \frac{4}{\varepsilon}(1 - e^{-w}) - \beta \int_{-\infty}^{x} w_y H w_y \, dy = 0. \tag{3.251}$$

Integrating (3.251) once from $-\infty$ to ∞ yields

$$\left(\frac{4}{\varepsilon} - a\right)\int_{-\infty}^{\infty} w \, dx - \frac{4}{\varepsilon}\int_{-\infty}^{\infty} (1 - e^{-w}) \, dx$$
$$- \beta \int_{-\infty}^{\infty} dx \int_{-\infty}^{x} w_y H w_y \, dy = 0. \tag{3.252}$$

Substituting the relation

$$\frac{1}{\varepsilon}\int_{-\infty}^{\infty} (1 - e^{-w}) \, dx = \int_{-\infty}^{\infty} u_s \, dx = \pi\beta, \tag{3.253}$$

which derives from (3.173) and (3.3), into (3.252) yields

$$\int_{-\infty}^{\infty} w\,dx = \frac{\pi\beta\varepsilon}{1 - a\varepsilon/4} + \frac{\beta\varepsilon/4}{1 - a\varepsilon/4}\int_{-\infty}^{\infty} dx \int_{-\infty}^{x} w_y H w_y\,dy. \quad (3.254)$$

However, since the second term on the right-hand side of (3.254) vanishes due to formula (App. III.25), (3.254) becomes

$$\int_{-\infty}^{\infty} w\,dx = \frac{\pi\beta\varepsilon}{1 - a\varepsilon/4} = \pi\beta \sum_{n=1}^{\infty} \left(\frac{a}{4}\right)^{n-1} \varepsilon^n. \quad (3.255)$$

It follows from (3.255), (3.182), and (3.185) that

$$I_n[u_s(x, t)] = \pi\beta(a/4)^{n-1}, \qquad n = 1, 2, \ldots. \quad (3.256)$$

For pure N solitons [(3.9) with (3.36)–(3.38)], (3.256) can be replaced by

$$I_n[u(x, t)] = \pi\beta \sum_{j=1}^{N} (a_j/4)^{n-1}, \qquad n = 1, 2, \ldots, \quad (3.257)$$

since, as will be shown in Chapter 4, $u(x, t)$ is expressed as a superposition of N separated solitons in the limit of large values of t. Using the fact that I_n is a conserved quantity

$$I_n[u(x, t)] = I_n[u(x, 0)], \qquad n = 1, 2, \ldots, \quad (3.258)$$

from (3.257), (3.258), and (3.185) we find that

$$\pi\beta \sum_{j=1}^{N} (a_j/4)^{n-1} = \int_{-\infty}^{\infty} w_n[u(x, 0)]\,dx, \qquad n = 1, 2, \ldots. \quad (3.259)$$

The first $N + 1$ expressions of (3.259) may be used to determine $a_1, a_2, \ldots, a_N$. The equations given by (3.259) can be rewritten in terms of y_j and s_n, defined by (3.239) and (3.241), as

$$\sum_{j=1}^{N} y_j^{n-1} = s_n, \qquad n = 1, 2, \ldots, N + 1. \quad (3.260)$$

For $n = 1$, Eq. (3.260) gives the total number N of solitons that arise from a given initial condition

$$N = s_1 = \frac{\sigma}{\pi}\int_{-\infty}^{\infty} \phi(\xi)\,d\xi. \quad (3.261)$$

For $2 \leq n \leq N + 1$, the equations defined by (3.260) give N equations for N unknown functions $y_1, y_2, \ldots, y_N$ from which the amplitudes of each soliton can be determined. To show this, it is convenient to introduce the elementary symmetric functions of $y_1, y_2, \ldots, y_N$ as

$$\sigma_1 = y_1 + y_2 + \cdots + y_N, \tag{3.262}$$

$$\sigma_2 = y_1 y_2 + y_2 y_3 + \cdots + y_{N-1} y_N, \tag{3.263}$$

$$\vdots$$

$$\sigma_{N-1} = y_1 y_2 \cdots y_{N-1} + \cdots + y_2 y_3 \cdots y_N, \tag{3.264}$$

$$\sigma_N = y_1 y_2 \cdots y_N. \tag{3.265}$$

The unknown quantities $y_1, y_2, \ldots, y_N$ are given by the roots of the single algebraic equation of order N [111]

$$y^N - \sigma_1 y^{N-1} + \sigma_2 y^{N-2} + \cdots + (-1)^N \sigma_N = 0. \tag{3.266}$$

The coefficients $\sigma_1, \sigma_2, \ldots, \sigma_N$ are uniquely determined by a recursion formula

$$(-1)^m m\sigma_m + (-1)^{m-1} s_2 \sigma_{m-1} + \cdots + s_{m-1}\sigma^2 - s_m \sigma_1 + s_{m+1} = 0, \quad m = 2, 3, \ldots, N, \tag{3.267}$$

starting with $\sigma_1 = s_2$. The first few terms σ_m are given as

$$\sigma_1 = s_2, \tag{3.268}$$

$$\sigma_2 = \tfrac{1}{2}(-s_3 + s_2^2), \tag{3.269}$$

$$\sigma_3 = \tfrac{1}{3}(s_4 - \tfrac{3}{2} s_2 s_3 + \tfrac{1}{2} s_2^3), \tag{3.270}$$

$$\sigma_4 = \tfrac{1}{4}(-s_5 + \tfrac{4}{3} s_2 s_4 + \tfrac{1}{2} s_3^2 - s_2^2 s_3 + \tfrac{1}{6} s_2^4). \tag{3.271}$$

The structure of the BO equation may be clarified by studying the properties of the algebraic equation (3.266).

3.3.2 *Asymptotic Solution in Zero Dispersion Limit*

We shall now investigate the asymptotic behavior of solutions when the effect of dispersion is very small, that is, $\beta \simeq 0$. In this case, the number of solitons arising from an initial condition $\phi(\xi)$ increases indefinitely, as seen from (3.261). Therefore, we can introduce the number density function $F(a)$ of solitons where $F(a)\,da$ gives the number dN of solitons with amplitudes within the interval $(a, a + da)$ as

$$dN = F(a)\,da. \tag{3.272}$$

Equation (3.259) is then approximated by the integral equations

$$\pi\beta \int_0^{\infty} \left(\frac{a}{4}\right)^{n-1} F(a)\,da = \frac{1}{n}\int_{-\infty}^{\infty} u^n(x, 0)\,dx$$
$$= \frac{u_0^n l}{n}\int_{-\infty}^{\infty} \phi^n(\xi)\,d\xi, \qquad \beta \to 0, \quad n = 1, 2, \ldots, \tag{3.273}$$

where we have used (3.191). The solution of (3.273) can be found to be [110, 111][§]

$$F(a) = \frac{\sigma}{4\pi u_0}\int_{4u_0\phi(\xi) > a} d\xi, \tag{3.274}$$

where the integration interval is restricted by the condition

$$4u_0\,\phi(\xi) > a. \tag{3.275}$$

It can be seen from (3.274) that the amplitudes of solitons do not exceed four times the maximum of the initial perturbation (3.234)

$$F(a) = 0 \qquad \text{for} \quad a > \max 4u(x, 0). \tag{3.276}$$

We now give explicit examples of $F(a)$ for two different initial conditions.

(i) Rectangular well

$$\phi(\xi) = \begin{cases} 1 & \text{for} \quad -\tfrac{1}{2} \le \xi \le \tfrac{1}{2}, & (3.277a) \\ 0 & \text{for} \quad \xi > \tfrac{1}{2}, \quad \xi < -\tfrac{1}{2}, & (3.277b) \end{cases}$$

$$F(a) = \begin{cases} \sigma/4\pi u_0 & \text{for} \quad a < 4u_0, & (3.278a) \\ 0 & \text{for} \quad a \ge 4u_0, & (3.278b) \end{cases}$$

$$N = \sigma/\pi; \tag{3.279}$$

[§] The proof of (3.274) is given in Appendix IV.

(ii) Lorentzian profile

$$\phi(\xi) = 1/(\xi^2 + 1), \tag{3.280}$$

$$F(a) = \begin{cases} (\sigma/2\pi u_0)(4u_0/a - 1)^{1/2} & \text{for} \quad a < 4u_0, \quad (3.281a) \\ 0 & \text{for} \quad a \geq 4u_0, \quad (3.281b) \end{cases}$$

$$N = \sigma. \tag{3.282}$$

The space distribution of solitons for large values of time is also obtained using (3.274). Since the individual soliton with amplitude a moves with velocity a and is found at $x = at$ when $t \to \infty$, we have from (3.275)

$$a = x/t, \qquad 0 < x/t < \max 4u(x, 0). \tag{3.283}$$

Denoting the number of solitons within the space interval $(x, x + dx)$ as $k(x, t)\,dx$, we obtain

$$k(x, t)\,dx = F(a)\,da, \tag{3.284}$$

which, combined with (3.283), yields

$$k(x, t) = (1/t)F(x/t), \tag{3.285}$$

where F is given by (3.274).

So far we have been concerned only with the initial condition defined in (3.234). However, these results can be generalized for the initial condition including the negative region where $u(x, 0) < 0$. In this case, after the lapse of a large amount of time, the initial disturbance is assumed to decay into separated N solitons with the amplitudes a_j ($j = 1, 2, \ldots, N$) and the tail part. (The tail part is a sort of dispersive wave.) The value of the conserved quantity may be assumed to be the sum of the I_n for the solitons and the I_n for the tail. Under these assumptions, the equations given by (2.259) can be replaced by

$$\pi\beta \sum_{j=1}^{N} (a_j/4)^{n-1} + \tilde{I}_n = \int_{-\infty}^{\infty} w_n(x, 0)\,dx, \qquad n = 1, 2, \ldots, \tag{3.286}$$

where $\tilde{I}_n$ denotes the value of $I_n[u(x, t)]$ for the tail. For small β, (3.286) is approximated by

$$\pi\beta \int_0^{\infty} (a/4)^{n-1} F(a)\,da + \tilde{I}_n = 1/n \int_{-\infty}^{\infty} u^n(x, 0)\,dx, \qquad n = 1, 2, \ldots. \tag{3.287}$$

Introducing $F(a)$ from (3.274) yields

$$1/n \int_{u(x,0)>0} u^n(x, 0)\, dx + \tilde{I}_n = 1/n \int_{-\infty}^{\infty} u^n(x, 0)\, dx, \qquad (3.288)$$

that is

$$\tilde{I}_n = 1/n \int_{u(x,0)<0} u^n(x, 0)\, dx, \qquad \beta \to 0, \qquad (3.289)$$

which implies that, for small β, the tail arises from the negative region of the initial disturbance $u(x, 0)$. The total number N of solitons arising from $\phi(\xi)$ is given in the limit of $\beta \to 0$ as

$$N = \int_0^{\infty} F(a)\, da = \sigma/\pi \int_{\phi(\xi)>0} \phi(\xi)\, d\xi. \qquad (3.290)$$

Hence, the number of solitons depends on the positive region of the initial disturbance, and for small β (or large σ) at least one soliton arises if the initial disturbance satisfies the condition

$$\int_{-\infty}^{\infty} \phi(\xi)\, d\xi > 0, \qquad (3.291)$$

as can be seen from (3.290).

3.3.3 *An Initial Condition Evolving into Pure N Solitons*

The initial conditions we have considered so far have been rather general. We shall now present an initial condition which evolves into pure N solitons after the lapse of a large amount of time. It has the form

$$u(x, 0) = \frac{N}{(x/\beta)^2 + 1}, \qquad N\text{: positive integer.} \qquad (3.292)$$

The amplitude of each soliton corresponding to (3.292) is identified with the zeros of the Laguerre polynomial of order N. To prove these statements, let us first define the Laguerre polynomial of order N by

$$L_N(y) = \sum_{r=0}^{N} (-1)^r \frac{N!}{(N-r)!(r!)^2} y^r. \qquad (3.293)$$

The $L_N(y)$ satisfies the following second-order ordinary differential equation

$$yL_N'' + (1 - y)L_N' + NL_N = 0, \tag{3.294}$$

where the prime denotes differentiation with respect to y.

Let y_j ($j = 1, 2, \ldots, N$) be N zeros of $L_N(y)$. Then the y_j have the properties

$$L_N(y_j) = 0, \qquad j = 1, 2, \ldots, N, \tag{3.295}$$

$$y_j > 0, \quad y_j \neq y_k \quad \text{for} \quad j \neq k, j, k = 1, 2, \ldots, N, \tag{3.296}$$

$$\sum_{\substack{k=1 \\ (k \neq j)}}^{N} (y_j - y_k)^{-1} = \tfrac{1}{2}(1 - y_j^{-1}), \qquad j = 1, 2, \ldots, N. \tag{3.297}$$

To show (3.297)[§], put $L_N(y)$ in the form

$$L_N(y) = (-1)^N \frac{1}{N!} \prod_{j=1}^{N} (y - y_j). \tag{3.298}$$

It follows by direct calculation that

$$L_N' = L_N \sum_{j=1}^{N} (y - y_j)^{-1}, \tag{3.299}$$

$$L_N'' = 2L_N \sum_{j=1}^{N} (y - y_j)^{-1} \sum_{\substack{k=1 \\ (k \neq j)}}^{N} (y_j - y_k)^{-1}. \tag{3.300}$$

Substituting (3.299) and (3.300) into (3.294) and rearranging, we obtain

$$\sum_{j=1}^{N} (y - y_j)^{-1} \left[\sum_{\substack{k=1 \\ (k \neq j)}}^{N} 2y_j(y_j - y_k)^{-1} + 1 - y_j \right] = 0. \tag{3.301}$$

Relation (3.301) holds for arbitrary y, and therefore the quantity in the parentheses must be zero identically, which implies (3.297).

We now define the matrix A whose elements are given by

$$A_{jk} = \delta_{jk}(z + y_j^{-1}) + 2(1 - \delta_{jk})(y_j - y_k)^{-1}, \tag{3.302}$$

[§] For a discussion of properties (3.295) and (3.296) see [113].

where z is an arbitrary complex number. As will be shown in Appendix II [see (App. II.4)], the following remarkable identity holds:

$$\det A(z) = (z + 1)^N. \tag{3.303}$$

Given these conditions, we employ the N-soliton solution of the BO equation [(3.9) with (3.36)–(3.38)]. Putting

$$a_j = y_j, \qquad j = 1, 2, \ldots, N, \tag{3.304a}$$

$$x_{0j} = 0, \qquad j = 1, 2, \ldots, N. \tag{3.304b}$$

in this solution, then at $t = 0$, the solution becomes [112]

$$\begin{aligned} u(x, 0) &= \frac{i\beta}{2} \frac{\partial}{\partial x} \ln \frac{f^*(x, 0)}{f(x, 0)} \\ &= \frac{i\beta}{2} \frac{\partial}{\partial x} \ln \frac{\det A^*(ix/\beta)}{\det A(ix/\beta)} \\ &= \frac{i\beta}{2} \frac{\partial}{\partial x} \ln \frac{(-ix/\beta + 1)^N}{(ix/\beta + 1)^N} \\ &= \frac{N}{(x/\beta)^2 + 1}, \end{aligned} \tag{3.305}$$

where we have used (3.303) in passing to the third line of (3.305). The asymptotic form of $u(x, t)$ corresponding to (3.305) for large values of time is expressed as

$$u(x, t) \underset{t \to \infty}{\simeq} \sum_{j=1}^{N} \frac{y_j}{(y_j/\beta)^2 (x - y_j t)^2 + 1}, \tag{3.306}$$

since the initial condition (3.305) evolves as $t \to \infty$ into the pure N solitons with amplitudes equal to the zeros y_j ($j = 1, 2, \ldots, N$) of the Laguerre polynomial of order N. The initial condition (3.292) has the same form as (3.125), which has already been derived from the linear algebraic equation.

We shall now evaluate the nth conserved quantity I_n ($n = 1, 2, \ldots$) of the BO equation for the initial condition (3.292). For this purpose, (3.257), (3.258), and (3.304) are used to obtain

$$I_n[u(x, 0)] = I_n[u(x, t)] = \pi\beta \sum_{j=1}^{N} \left(\frac{a_j}{4}\right)^{n-1} = \frac{\pi\beta}{4^{n-1}} \sum_{j=1}^{N} y_j^{n-1}. \tag{3.307}$$

Therefore, it is sufficient to calculate the quantity

$$p_n \equiv \sum_{j=1}^{N} y_j^n, \qquad n = 1, 2, \ldots, \tag{3.308}$$

the solution of which is given by the well-known Euler formula as

$$p_n = \begin{cases} \sigma_1 p_{n-1} - \sigma_2 p_{n-2} + \cdots + (-1)^{n-2}\sigma_{n-1}p_1 + (-1)^{n-1}n\sigma_n \\ \qquad\qquad \text{for} \quad n = 1, 2, \ldots, N, & (3.309) \\ \sigma_1 p_{n-1} - \sigma_2 p_{n-2} + \cdots + (-1)^{n-1}\sigma_N p_{n-N} \\ \qquad\qquad \text{for} \quad n > N. & (3.310) \end{cases}$$

Here $\sigma_1, \sigma_2, \ldots, \sigma_N$ are elementary symmetric functions of $y_1, y_2, \ldots, y_N$ defined in (3.262)–(3.265). In this case, by comparing (3.293) and (3.298), it follows that

$$\sigma_j = [N!/(N-j)!]^2/j!, \qquad j = 1, 2, \ldots, N. \tag{3.311}$$

The p_n $(n = 1, 2, \ldots)$ are calculated by introducing (3.311) into (3.309) and (3.310). The first few expressions of p_n are given as

$$p_1 = N^2, \tag{3.312}$$

$$p_2 = N^2(2N - 1), \tag{3.313}$$

$$p_3 = N^2(5N^2 - 6N + 2), \tag{3.314}$$

$$p_4 = N^2(14N^3 - 29N^2 + 22N - 6), \tag{3.315}$$

$$p_5 = N^2(42N^4 - 130N^3 + 165N^2 - 100N + 24). \tag{3.316}$$

In particular, for $N = 1$,

$$p_n = 1, \tag{3.317}$$

so that

$$I_n\{1/[(x/\beta)^2 + 1]\} = \pi\beta/4^{n-1}. \tag{3.318}$$

Finally, we may note a close relationship between the distribution function $F(a)$ of solitons and that of the zeros of the Laguerre polynomial of order N in the limit of large N. Since the amplitudes of the solitons are given by the zeros of L_N for the initial condition (3.292), we obtain, by substituting (3.292) into (3.274) and setting $a = y$, in the limit of $N \to \infty$

$$F(y) = \begin{cases} (1/2\pi)(4N/y - 1)^{1/2}, & y < 4N, \quad (3.319) \\ 0, & y \geq 4N. \quad (3.320) \end{cases}$$

Here $F(y)$ is normalized such that the total number of zeros corresponds to N:

$$\int_0^\infty F(y)\,dy = N. \tag{3.321}$$

The $F(y)\,dy$ gives the number of zeros within the interval $(y, y + dy)$. Using (3.319) and (3.320) the p_n defined in (3.308) are evaluated in the limit of $N \to \infty$ as

$$\begin{aligned} p_n &= \sum_{j=1}^{N} y_j^n \\ &\simeq \int_0^\infty y^n F(y)\,dy \\ &= \frac{1}{2\pi}\int_0^{4N} y^n\left(\frac{4N}{y} - 1\right)^{1/2} dy \\ &= \frac{(2n)!}{n!(n+1)!} N^{n+1}, \qquad n = 1, 2, \ldots, \end{aligned} \tag{3.322}$$

which correspond to the first term on the right-hand side of (3.312)–(3.316).

3.4 Stability of the Benjamin–Ono Solitons

In this section the linear stability problem of the BO solitons is briefly discussed using the results obtained in Section 3.3. Let us consider the initial condition $u(x, 0)$ which evolves completely into pure N solitons after the lapse of a large amount of time. The amplitude of each soliton is then determined by the system of equations (3.259) as

$$\pi\beta \sum_{j=1}^{N} (a_j/4)^n = \int_{-\infty}^{\infty} w_{n+1}[u(x, 0)]\,dx = I_{n+1}, \qquad n = 1, 2, \ldots. \tag{3.323}$$

Consider the increments δa_j of the amplitude of each soliton when the initial disturbance $u(x, 0)$ is varied infinitesimally, by $\delta u(x, 0)$, for example. It follows from (3.323) that

$$\frac{\pi\beta n}{4^n} \sum_{i=1}^{N} a_j^{n-1}\,\delta a_j = \delta I_{n+1}, \qquad n = 1, 2, \ldots, N, \tag{3.324}$$

where δI_{n+1} denotes the variation of I_n ($n = 1, 2, \ldots, N$) corresponding to $\delta u(x, 0)$.

The system of equations (3.324) is linear with respect to δa_j, and the solutions are obtained using Cramer's formula as

$$\delta a_j = \begin{vmatrix} 1 & 1 & \overbrace{4\,\delta I_2/\pi\beta}^{j\text{th column}} & \cdots & 1 \\ a_1 & a_2 & 8\,\delta I_3/\pi\beta & \cdots & a_N \\ \vdots & \vdots & \vdots & \ddots & \vdots \\ a_1^{N-1} & a_2^{N-1} & \dfrac{4^N}{N}\,\delta I_{N+1}/\pi\beta & \cdots & a_N^{N-1} \end{vmatrix} \times \begin{vmatrix} 1 & 1 & \cdots & 1 \\ a_1 & a_2 & \cdots & a_N \\ \vdots & \vdots & \ddots & \vdots \\ a_1^{N-1} & a_2^{N-1} & \cdots & a_N^{N-1} \end{vmatrix}^{-1}, \quad j = 1, 2, \ldots, N. \tag{3.325}$$

Here the denominator of (3.325) is the Vandermonde determinant

$$\begin{vmatrix} 1 & 1 & \cdots & 1 \\ a_1 & a_2 & \cdots & a_N \\ \vdots & \vdots & \ddots & \vdots \\ a_1^{N-1} & a_2^{N-1} & \cdots & a_N^{N-1} \end{vmatrix} = \prod_{1 \le k < j \le N} (a_j - a_k). \tag{3.326}$$

It may be seen from (3.325) and (3.326) that a small variation in the initial disturbance results in a small change in the amplitude of each soliton since $a_j \neq a_k$ for $j \neq k$. This means that the N-soliton solution is stable against the small disturbance.

3.5 The Linearized Benjamin–Ono Equation and Its Solution

In this section the linearized BO equation is analyzed using the Fourier transform method. The initial value problem of the linearized BO equation is solved exactly, and the asymptotic behavior of the solution is investigated for large values of time. The self-similar solution of the BO equation is also discussed briefly.

Consider the linearized version of the BO equation in the form

$$u_t + Hu_{xx} = 0. \tag{3.327}$$

If $u(x, t)$ is represented in the form of the Fourier transform as

$$u(x, t) = \int_{-\infty}^{\infty} v(k) \exp\{i[kx - \omega(k)t]\}\, dk, \tag{3.328}$$

where k and $\omega(k)$ are the wave number and frequency, respectively, then substituting (3.328) into (3.327) yields the dispersion relation

$$\omega(k) = -k|k|, \tag{3.329}$$

where use has been made of a formula

$$P \int_{-\infty}^{\infty} \frac{e^{ikx}}{x}\, dx = i\pi \frac{|k|}{k}. \tag{3.330}$$

The unknown function $v(k)$ is determined from the initial value $u(x, 0)$ as

$$v(k) = \frac{1}{2\pi} \int_{-\infty}^{\infty} u(x, 0) e^{-ikx}\, dx. \tag{3.331}$$

Substituting (3.331) and (3.329) into (3.328), we obtain a general solution of (3.327) as

$$\begin{aligned} u(x, t) &= \frac{1}{2\pi} \int_{-\infty}^{\infty} \int_{-\infty}^{\infty} \exp\{i[k(x - y) + k|k|t]\} u(y, 0)\, dk\, dy \\ &= \frac{1}{\sqrt{2\pi t}} \int_{-\infty}^{\infty} K\left(\frac{x - y}{2\sqrt{t}}\right) u(y, 0)\, dy, \end{aligned} \tag{3.332}$$

where

$$\begin{aligned} K(x) &= 1/\sqrt{2\pi} \int_{-\infty}^{\infty} \exp[i(2kx + k|k|)]\, dk \\ &= [\tfrac{1}{2} - C(\sqrt{2/\pi}\,x)] \cos(x^2) + [\tfrac{1}{2} - S(\sqrt{2/\pi}\,x)] \sin(x^2), \end{aligned} \tag{3.333}$$

with

$$C(x) = \int_0^x \cos(\pi t^2/2)\, dt, \tag{3.334}$$

$$S(x) = \int_0^x \sin(\pi t^2/2)\, dt. \tag{3.335}$$

The function $K(x)$ defined in (3.333) has the following asymptotic expansions:

$$K(x) \simeq \begin{cases} \sqrt{\dfrac{2}{\pi}} \displaystyle\sum_{n=0}^{\infty} \dfrac{(-1)^n(4n+1)!!}{2^{2n+2}x^{4n+3}}, & x \to +\infty, \quad (3.336) \\ \cos(x^2) + \sin(x^2) - \sqrt{\dfrac{2}{\pi}} \displaystyle\sum_{n=0}^{\infty} \dfrac{(-1)^n(4n+1)!!}{2^{2n+2}x^{4n+3}}, & x \to -\infty. \quad (3.337) \end{cases}$$

Now assume that the initial condition vanishes rapidly when $|x| \to \infty$. Expanding $K[(x-y)/2\sqrt{t}]$ in powers of y as

$$K\left(\frac{x-y}{2\sqrt{t}}\right) = \sum_{n=0}^{\infty} \frac{(-1)^n}{n!} K^{(n)}\left(\frac{x}{2\sqrt{t}}\right) y^n, \tag{3.338}$$

with

$$K^{(n)}(x) = d^n K(x)/dx^n, \tag{3.339}$$

and inserting (3.338) into (3.332), we find

$$u(x, t) = \frac{1}{\sqrt{2\pi t}} \sum_{n=0}^{\infty} \frac{(-1)^n P_n}{n!} (2\sqrt{t})^{-n} K^{(n)}\left(\frac{x}{2\sqrt{t}}\right), \tag{3.340}$$

where

$$P_n = \int_{-\infty}^{\infty} y^n u(y, 0)\, dy, \qquad n = 0, 1, \ldots. \tag{3.341}$$

The asymptotic expressions for $K^{(n)}$ follow from (3.336) and (3.337) as

$$K^{(n)}(x) \simeq \begin{cases} \sqrt{2/\pi}[(-1)^n/8][(n+2)!/x^{n+3}], & x \to +\infty \quad (3.342) \\ \sqrt{2}(2x)^n \cos(x^2 - \pi/4 + n\pi/2), & x \to -\infty, \quad (3.343) \end{cases}$$

and the Fourier transform of $u(x, 0)$ is represented, by using (3.331) and (3.341), as

$$v(k) = \frac{1}{2\pi} \sum_{n=0}^{\infty} P_n \frac{(-ik)^n}{n!}. \tag{3.344}$$

The asymptotic expression of $u(x, t)$ is derived by employing the above results as follows: When $x/\sqrt{t} \to -\infty$, we find from (3.340) and (3.343) that

$$u(x, t) \simeq 2\sqrt{\pi/t}\, \mathrm{Re}\{v(x/2t) \exp[i(x^2/4t - \pi/4)]\}. \tag{3.345}$$

When $x/\sqrt{t} \to +\infty$, it follows from (3.340) and (3.342) that

$$u(x, t) \simeq (1/\pi\sqrt{t})[2/(x/\sqrt{t})^3]P_0, \qquad P_0 \neq 0. \tag{3.346}$$

For $|x|/\sqrt{t} \ll 1$ and large values of t, the solution is approximated by the first nonvanishing term in (3.340) to yield

$$u(x, t) \simeq (1/\sqrt{2\pi t})K(x/2\sqrt{t})P_0, \qquad P_0 \neq 0. \tag{3.347}$$

As an explicit example of the initial condition, consider Dirac's delta function

$$u(x, 0) = \delta(x). \tag{3.348}$$

The solution for $t > 0$ is given by (3.332) as

$$u(x, t) = (1/\sqrt{2\pi t})K(x/2\sqrt{t}), \tag{3.349}$$

and asymptotic expressions of (3.349) are found from (3.336) and (3.337) as

$$u(x, t) \simeq \begin{cases} (1/\pi\sqrt{t})[1/(x/\sqrt{t})^3] & \text{for } x/\sqrt{t} \to +\infty, \quad (3.350) \\ (1/\sqrt{\pi t}) \cos(x^2/4t - \pi/4) & \text{for } x/\sqrt{t} \to -\infty. \quad (3.351) \end{cases}$$

Finally, we shall comment on the similarity solution of the BO equation. It may easily be confirmed that the BO equation (3.1) is invariant under the similarity transformations

$$u = u'/\sqrt{\varepsilon}, \tag{3.352}$$

$$x = x'/\sqrt{\varepsilon}, \tag{3.353}$$

$$t = \varepsilon t'. \tag{3.354}$$

Therefore, u has a solution in the form

$$u = (1/\sqrt{t})f(x/\sqrt{t}). \tag{3.355}$$

Substituting (3.355) into the BO equation (3.1) yields

$$-\tfrac{1}{2}(\xi f' + f) + 4ff' + Hf'' = 0, \tag{3.356}$$

where the prime denotes differentiation with respect to the similarity variable ξ;

$$\xi = x/\sqrt{t}. \tag{3.357}$$

If the boundary condition

$$f(+\infty) = 0 \tag{3.358}$$

is imposed, (3.356) reduces, after integration by ξ, to

$$-\tfrac{1}{2}\xi f + 2f^2 + Hf' = 0. \tag{3.359}$$

It should be noted that the linearized version of (3.359), that is,

$$-\tfrac{1}{2}\xi f + Hf' = 0, \tag{3.360}$$

has a solution

$$f(\xi) = K(\xi/2^{4/3}), \tag{3.361}$$

where K is given by (3.333). The analytical solution of (3.359) has not yet been obtained, therefore the numerical method may be applied to it.

4

Interaction of the Benjamin–Ono Solitons

In this chapter we study the properties of the N-soliton solution of the BO equation. The asymptotic behavior of the solution is derived for large value of time, and it is shown that no phase shift appears as a result of soliton collisions.[§] The very complex interaction process between two solitons is then investigated in detail. The motion of poles corresponding to the two-soliton solution is drawn in a complex plane, and the nature of the interaction is clarified.[¶]

[§] This is in contrast to collisions that take place between KdV solitons.

[¶] The main part of this chapter follows the discussion of Matsuno [114].

4.1 Asymptotic Behaviors of the N-Soliton Solution

Consider the N-soliton solution of the BO equation in the form expressed by (3.9) and (3.36)–(3.38), and expand the determinant with respect to the variables θ_j ($j = 1, 2, \ldots, N$) as

$$f_N = \sum_{n=1}^{N} \sum_{{}_NC_n} i^n Q_{N-n}(s_1, s_2, \ldots, s_n)\theta_{s_1}\theta_{s_2} \cdots \theta_{s_n}$$

$$= i^N\theta_1\theta_2 \cdots \theta_N + i^{N-1}\sum_{n=1}^{N} \frac{1}{a_n}\theta_1\theta_2 \cdots \theta_{n-1}\theta_{n+1} \cdots \theta_N + \cdots, \tag{4.1}$$

with

$$Q_{N-n}(s_1, s_2, \ldots, s_n) = \frac{1}{i^n}\left(\frac{\partial^n}{\partial\theta_{s_1} \cdots \partial\theta_{s_n}} f_N\right)\theta_1 = \theta_2 = \cdots = \theta_N = 0,$$

$$n = 1, 2, \ldots, N-1, \tag{4.2}$$

$$Q_N = [f_N]\theta_1 = \theta_2 = \cdots = \theta_N = 0, \tag{4.3}$$

$$Q_0 = 1, \tag{4.4}$$

where the notation $\sum_{{}_NC_n}$ means summation over all possible combinations of n elements taken from $1, 2, \ldots, N$.

To investigate the behavior of (4.1), we order the velocities of each soliton as

$$0 < a_1 < a_2 < \cdots < a_N, \tag{4.5}$$

transform to a moving frame with a velocity a_n, and then take the limits $t \to \pm\infty$. First, consider the limit $t \to -\infty$. It follows from (3.8) and (4.1) that

$$\theta_1, \theta_2, \ldots, \theta_{n-1} = -\infty, \tag{4.6}$$

$$\theta_n = \text{finite}, \tag{4.7}$$

$$\theta_{n+1}, \theta_{n+2}, \ldots, \theta_N = +\infty. \tag{4.8}$$

Substituting (4.6)–(4.8) into (4.1), we obtain the asymptotic behavior of f_N as

$$f_N = i^N\theta_1\theta_2 \cdots \theta_{n-1}\theta_n\theta_{n+1} \cdots \theta_N$$
$$+ i^{N-1}(1/a_n)\theta_1\theta_2 \cdots \theta_{n-1}\theta_{n+1} \cdots \theta_N + \cdots. \tag{4.9}$$

Thus, we find from (3.9) and (4.9) that

$$
\begin{aligned}
u &= \frac{i}{2}\frac{\partial}{\partial x}\ln\frac{f_N^*}{f_N} \\
&\simeq \frac{i}{2}\frac{\partial}{\partial x}\ln\Bigg\{\bigg[(-i)^N\theta_1\theta_2\cdots\theta_{n-1}\theta_n\theta_{n+1}\cdots\theta_N \\
&\quad + (-i)^{N-1}\frac{1}{a_n}\theta_1\theta_2\cdots\theta_{n-1}\theta_{n+1}\cdots\theta_N\bigg]\Big/ \\
&\quad \bigg[i^N\theta_1\theta_2\cdots\theta_{n-1}\theta_n\theta_{n+1}\cdots\theta_N + i^{N-1}\frac{1}{a_n}\theta_1\theta_2\cdots\theta_{n-1}\theta_{n+1}\cdots\theta_N\bigg]\Bigg\} \\
&= \frac{i}{2}\frac{\partial}{\partial x}\ln\left\{\left[(-1)^{N-1}\left(-i\theta_n + \frac{1}{a_n}\right)\right]\Big/\left(i\theta_n + \frac{1}{a_n}\right)\right\} \\
&= \frac{a_n}{(a_n\theta_n)^2 + 1}, \qquad t \to -\infty,
\end{aligned}
\tag{4.10}
$$

which is identical to the one-soliton solution (3.3) of amplitude a_n moving to the right at constant velocity a_n. The asymptotic behavior for $t \to +\infty$ is derived similarly. The result is the same as (4.10), that is,

$$
u \simeq \frac{a_n}{(a_n\theta_n)^2 + 1}, \qquad t \to +\infty. \tag{4.11}
$$

Therefore, the solution is found to evolve asymptotically as $t \to \pm\infty$ into N localized solitons moving with constant velocities $a_1, a_2, \ldots, a_N$ in the original reference frame as

$$
u(x, t) \simeq \sum_{n=1}^{N} \frac{a_n}{(a_n\theta_n)^2 + 1}, \qquad t \to \pm\infty. \tag{4.12}
$$

It can be seen from (4.12) that no phase shift appears as the result of collisions of solitions, in contrast to collisions that take place between the KdV solitons [see (2.64)]. This is an interesting feature of the system of solutions expressed by (3.9) and (3.36)–(3.38). In Fig. 4.1 (a)–(d) and 4.2 (a)–(d), the $u(x, t)$ for $N = 2$, 3 are plotted as a function of x for various values of t.

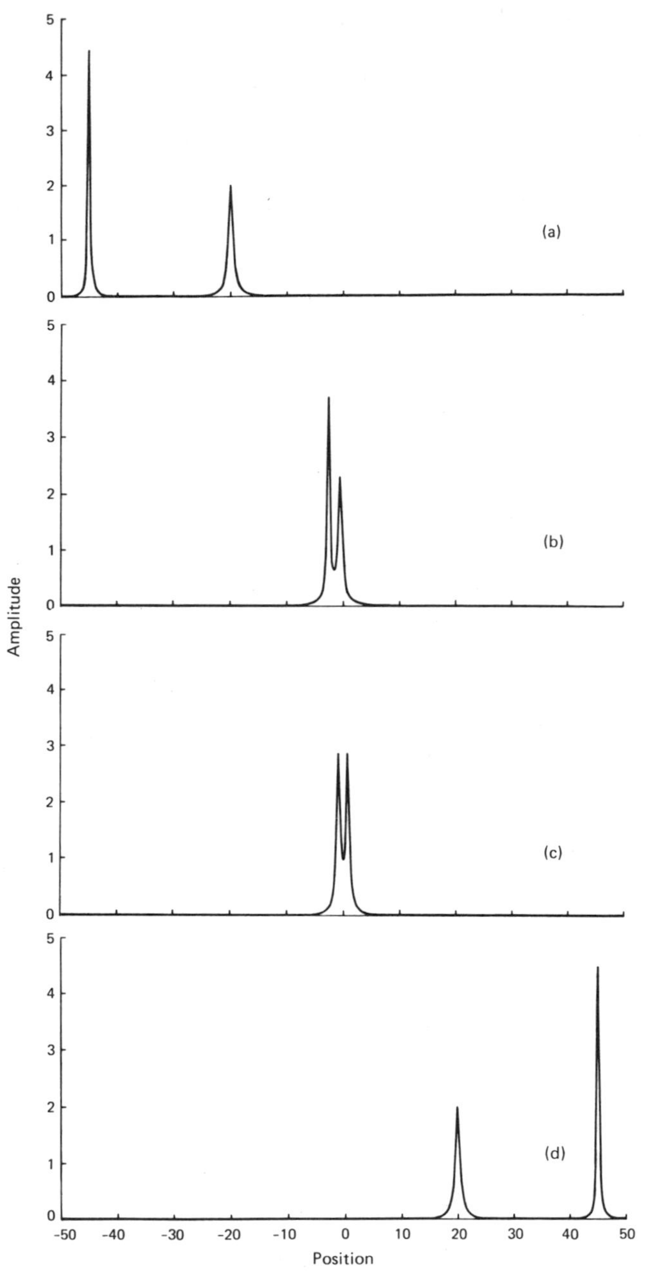

Fig. 4.1 Plot of $u(x, t)$ for $N = 2$, $a_1 = 2.0$, $a_2 = 4.5$, $x_{01} = x_{02} = 0$, and (a) $t = -10.0$, (b) $t = -0.05$, (c) $t = 0$, (d) $t = 10.0$.

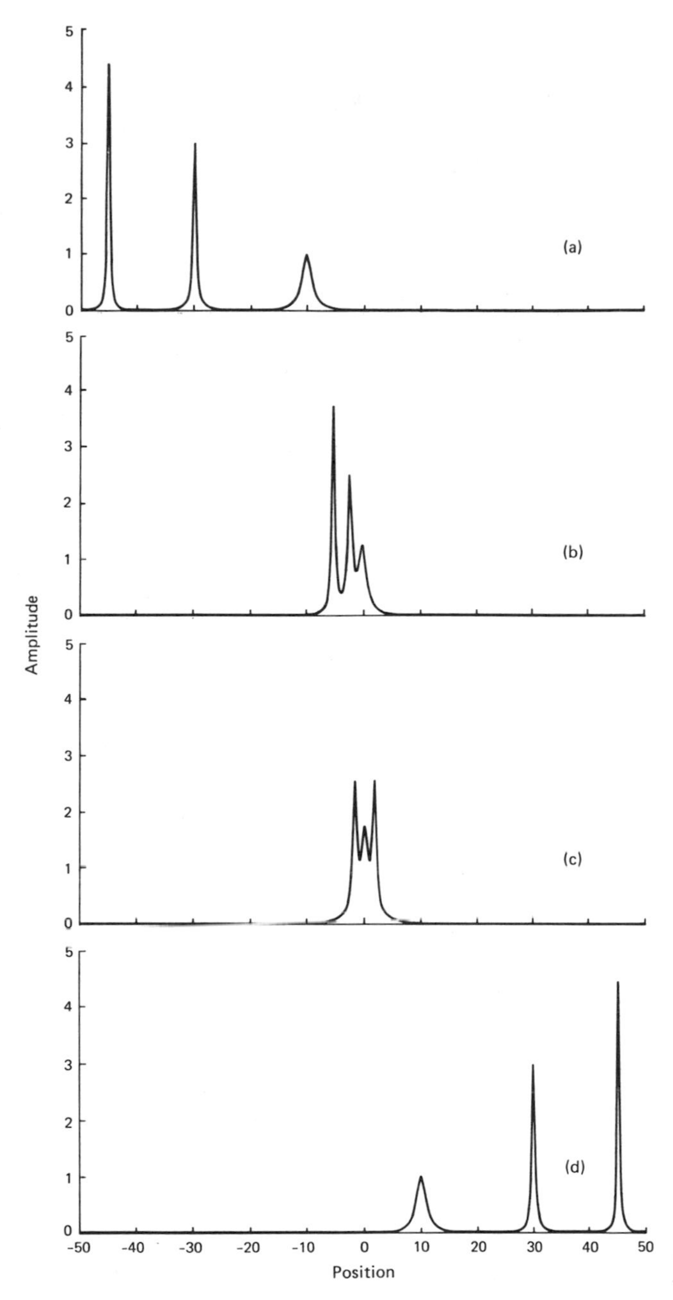

Fig. 4.2 Plot of $u(x, t)$ for $N = 3, a_1 = 1.0, a_2 = 3.0, a_3 = 4.5, x_{01} = x_{02} = x_{03} = 0$, and (a) $t = -10.0$, (b) $t = -1.0$, (c) $t = 0$, (*d*) $t = 10.0$.

We shall now derive the asymptotic form of $u(x, t)$ when $|x| \to \infty$ while keeping t finite. The asymptotic form of f_N for this limit can be obtained using expansion (4.1) as

$$f_N \simeq i^N x^N + i^{N-1}\left[\sum_{n=1}^{N} \frac{1}{a_n} - i \sum_{n=1}^{N} (a_n t + x_{0n})\right] x^{N-1} + \cdots. \tag{4.13}$$

Thus, we find

$$\begin{aligned} u(x, t) &= \frac{i}{2} \frac{\partial}{\partial x} \ln \frac{f_N^*}{f_N} \\ &= \frac{i}{2} \frac{\partial}{\partial x} \ln \frac{1 + i \sum_{n=1}^{N} [1/a_n + i(a_n t + x_{0n})](1/x) + \cdots}{1 - i \sum_{n=1}^{N} [1/a_n - i(a_n t + x_{0n})](1/x) + \cdots} \\ &= \sum_{n=1}^{N} \frac{1}{a_n} \frac{1}{x^2} + O(x^{-4}), \end{aligned} \tag{4.14}$$

which implies that $u(x, t)$ has a long tail, unlike the asymptotic behavior of the KdV N-soliton solution, which decays exponentially.

4.2 Interaction of Two Solitons

As shown in Section 4.1, the BO solitons are remarkable in that they cause no phase shift after interactions. However, during the interaction, the processes are very complex. Therefore, it is necessary to find a representation of solitons that brings to light more details of the interaction process. For this purpose, the interaction of two solitons is studied in this section. The motion of two solitons may be characterized by that of the poles $x_n(t)$ $(n = 1, 2)$ in a complex plane [see (3.9) and (3.10)]. To show this, we rewrite the N-solition solution (3.9) with (3.10) in the form

$$\begin{aligned} u(x, t) &= \frac{1}{2i}\left[\sum_{n=1}^{N} \frac{1}{x - x_n(t)} - \sum_{n=1}^{N} \frac{1}{x - x_n^*(t)}\right] \\ &= \sum_{n=1}^{N} \frac{\operatorname{Im} x_n(t)}{[x - \operatorname{Re} x_n(t)]^2 + [\operatorname{Im} x_n(t)]^2}. \end{aligned} \tag{4.15}$$

The term $u(x, t)$ has $2N$ poles $x_1(t), x_2(t), \ldots, x_N(t), x_1^*(t), x_2^*(t), \ldots, x_N^*(t)$ in the complex x plane. The trajectories of $x_n(t)$ $(n = 1, 2, \ldots, N)$ can be obtained directly from (3.10) and (3.36)–(3.38) by solving an algebraic equation of order N. Note that Re x_n gives the center position and $(\text{Im } x_n)^{-1}$ the amplitude of the nth soliton.

For the two-soliton case, poles $x_1(t)$ and $x_2(t)$ are found from (3.10) and (3.32) by solving the algebraic equation of order two

$$-\theta_1\theta_2 + i\left(\frac{\theta_1}{a_2} + \frac{\theta_2}{a_1}\right) + \frac{1}{a_1a_2}\left(\frac{a_1 + a_2}{a_1 - a_2}\right)^2 = 0. \tag{4.16}$$

The solutions of (4.16) are obtained as

$$x_1(t) = \frac{a_1 + a_2}{2}T + \frac{a(T)}{2} + \frac{a_1x_{02} - a_2x_{01}}{a_1a_2} + i\left[\frac{a_1 + a_2}{2a_1a_2} + \frac{b(T)}{2}\right], \tag{4.17}$$

$$x_2(t) = \frac{a_1 + a_2}{2}T - \frac{a(T)}{2} + \frac{a_1x_{02} - a_2x_{01}}{a_1a_2} + i\left[\frac{a_1 + a_2}{2a_1a_2} - \frac{b(T)}{2}\right], \tag{4.18}$$

where

$$a(T) = -\frac{(a_1 - a_2)^2}{a_1a_2}\frac{T}{b(T)}, \tag{4.19}$$

$$\begin{aligned} b(T) = \frac{1}{\sqrt{2}}\Bigg(-\left[(a_1 - a_2)^2T^2 - \left(\frac{a_1 + a_2}{a_1a_2}\right)^2 + \frac{4}{a_1a_2}\left(\frac{a_1 + a_2}{a_1 - a_2}\right)^2\right] \\ + \left\{\left[(a_1 - a_2)^2T^2 - \left(\frac{a_1 + a_2}{a_1a_2}\right)^2 + \frac{4}{a_1a_2}\left(\frac{a_1 + a_2}{a_1 - a_2}\right)^2\right]^2 \right. \\ \left. + 4\frac{(a_1 - a_2)^4}{(a_1a_2)^2}T^2\right\}^{1/2}\Bigg)^{1/2}, \end{aligned} \tag{4.20}$$

$$T = t + \frac{x_{01} - x_{02}}{a_1 - a_2}. \tag{4.21}$$

In the following discussion, phase constants x_{01} and x_{02} are assumed to be zero without loss of generality, so that

$$T = t. \tag{4.22}$$

In the limit of $t \to \pm\infty$, $a(t)$ and $b(t)$ behave like

$$a(t) = (a_1 - a_2)t + O(t^{-1}), \tag{4.23}$$

$$b(t) = (a_2 - a_1)/(a_1 a_2) + O(t^{-2}). \tag{4.24}$$

Substituting (4.23) and (4.24) into (4.17) and (4.18), we find, in the same limit, that

$$x_1(t) = a_1 t + \frac{i}{a_1} + O(t^{-1}), \tag{4.25}$$

$$x_2(t) = a_2 t + \frac{i}{a_2} + O(t^{-1}). \tag{4.26}$$

The above results may also be derived directly from (4.12) and (4.15) as

$$\begin{aligned} u(x, t) &\simeq \sum_{n=1}^{N} \frac{a_n}{(a_n \theta_n)^2 + 1} \\ &= \frac{1}{2i}\left(\sum_{n=1}^{N} \frac{1}{\theta_n - i a_n^{-1}} - \sum_{n=1}^{N} \frac{1}{\theta_n + i a_n^{-1}}\right) \\ &= \frac{1}{2i}\left[\sum_{n=1}^{N} \frac{1}{x - (a_n t + x_{0n} + i a_n^{-1})} \right. \\ &\quad \left. - \sum_{n=1}^{N} \frac{1}{x - (a_n t + x_{0n} - i a_n^{-1})}\right], \qquad t \to \pm\infty. \end{aligned} \tag{4.27}$$

If we regard solitons as stable particles located at the poles, we may introduce the time t_c at which two solitons collide. It is natural to define it such that the distance between two poles reduces to a minimum. In the present case, the distance $l(t)$ between two poles at time t is found from (4.17) and (4.18) as

$$l(t) = |x_1(t) - x_2(t)| = [a^2(t) + b^2(t)]^{1/2}. \tag{4.28}$$

It follows from (4.19) and (4.22) that

$$l(t) = \left[\frac{(a_1 - a_2)^4}{(a_1 a_2)^2} \frac{t^2}{b^2(t)} + b^2(t)\right]^{1/2} \geq \left[2 \frac{(a_1 - a_2)^2}{a_1 a_2} |t|\right]^{1/2}, \tag{4.29}$$

by the inequality

$$a + b \geq 2\sqrt{ab}, \qquad a, b \geq 0. \tag{4.30}$$

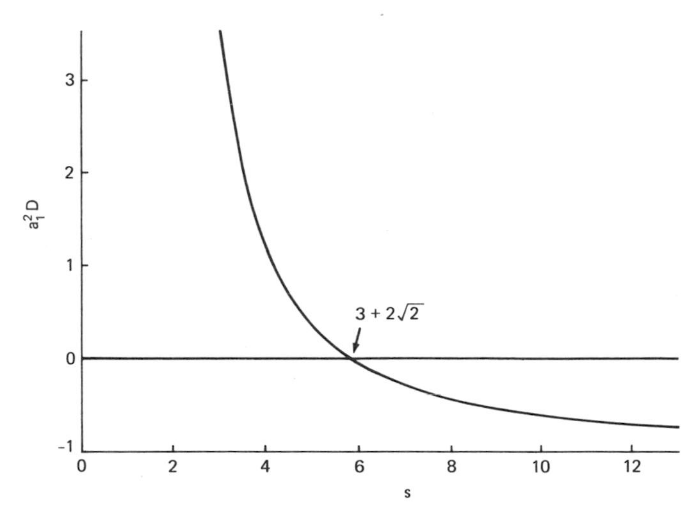

Fig. 4.3 Plot of $a_1^2 D$ as a function of s $(=a_2/a_1)$. (From reference [114], by permission of The Institute of Physics, England.)

It may be seen that $l(t)$ reduces to a minimum when $t = 0$. Therefore, it is especially important to investigate the behavior of poles in the situation near $t = 0$.

The nature of the interaction of poles $x_1(t)$ and $x_2(t)$ is characterized by the behavior of the function $b(t)$ as seen from (4.17)–(4.19). The behavior of $b(t)$ depends critically on the quantity

$$D \equiv \frac{4}{a_1 a_2}\left(\frac{a_1 + a_2}{a_1 - a_2}\right)^2 - \left(\frac{a_1 + a_2}{a_1 a_2}\right)^2 = \frac{1}{a_1^2}\frac{(s+1)^2}{s^2}\left[\frac{4s}{(s-1)^2} - 1\right], \tag{4.31}$$

where

$$s = a_2/a_1. \tag{4.32}$$

It will be shown later that D defined in (4.31) is merely the square of the minimum distance between two poles. Note that $a_1^2 D$ decreases monotonically from ∞ to -1 as s increases from 1 to ∞, as shown in Fig. 4.3.

We shall now study the behavior of $x_1(t)$ and $x_2(t)$ for small t under the conditions

$$D > 0, \qquad 1 < s < 3 + 2\sqrt{2}, \tag{4.33}$$

$$D = 0, \qquad s = 3 + 2\sqrt{2}, \tag{4.34}$$

$$D < 0, \qquad s > 3 + 2\sqrt{2}. \tag{4.35}$$

The discussion following is concerned only with the situation near $t = 0$ since we are interested in the details of the process of the collision of solitons.

Case A. When $D > 0$, condition (4.33), the behaviors of $a(t)$, $b(t)$, $x_1(t)$, and $x_2(t)$ for small t are

$$a(t) = (-\sqrt{D}t/|t|)[1 + O(t^2)], \tag{4.36}$$

$$b(t) = (|t|/\sqrt{D})[1 + O(t^2)], \tag{4.37}$$

$$x_1(t) = -\tfrac{1}{2}(\sqrt{D}t/|t|) + i(a_1 + a_2)/(a_1 a_2) + O(t), \tag{4.38}$$

$$x_2(t) = \tfrac{1}{2}(\sqrt{D}t/|t|) + i(a_1 + a_2)/(a_1 a_2) + O(t). \tag{4.39}$$

It may be seen from these expressions that Re $x_1(t)$ and Re $x_2(t)$ are discontinuous at $t = 0$. The minimum distance between two poles is given by

$$l(t_c) = l(0) = \sqrt{D}. \tag{4.40}$$

Figure 4.4 shows a plot of a two-soliton solution for various values of time, with parameters given by

$$a_1 = 1.0, \tag{4.41}$$

$$a_2 = 5.6, \tag{4.42}$$

$$x_{01} = x_{02} = 0, \tag{4.43}$$

$$D = 0.0814, \tag{4.44}$$

$$l(t_c) = l(0) = 0.285. \tag{4.45}$$

Figure 4.5 represents the motion of two poles for $-0.3 \leq t \leq 0.3$. In Fig. 4.6, Re x_1 and Re x_2 are plotted as functions of t. We can see from these figures that for $D > 0$ two poles interchange their velocities at the instant of the collision ($t = 0$), which reflects the discontinuities of Re x_1 and Re x_2 at $t = 0$ [see (4.38) and (4.39)].

Case B. When $D = 0$, condition (4.34),

$$a(t) = -(2t/|t|^{1/2})[1 + O(|t|^{3/2})], \tag{4.46}$$

$$b(t) = 2|t|^{1/2}[1 + O(|t|^{3/2})], \tag{4.47}$$

$$x_1(t) = -t/|t|^{1/2} + i(0.586/a_1 + |t|^{1/2}) + O(t), \tag{4.48}$$

$$x_2(t) = -t/|t|^{1/2} + i(0.586/a_1 - |t|^{1/2}) + O(t). \tag{4.49}$$

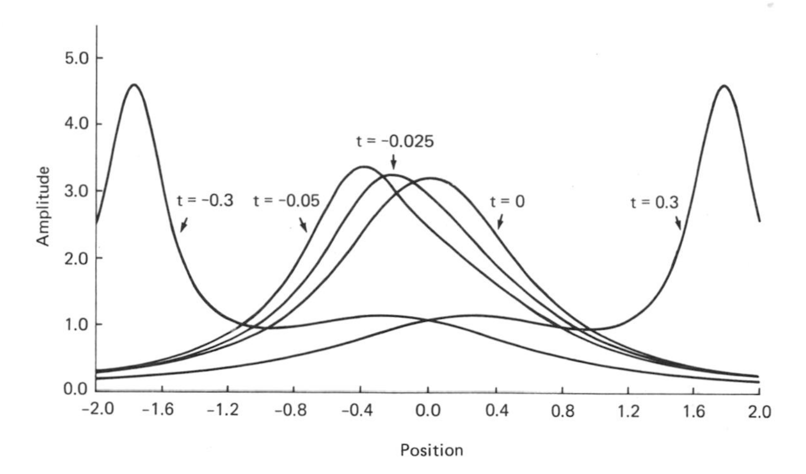

Fig. 4.4 Plot of a two-soliton solution with $a_1 = 1.0$, $a_2 = 5.6$, $x_{01} = x_{02} = 0$ for various values of time t. (From reference [114], by permission of The Institute of Physics, England.)

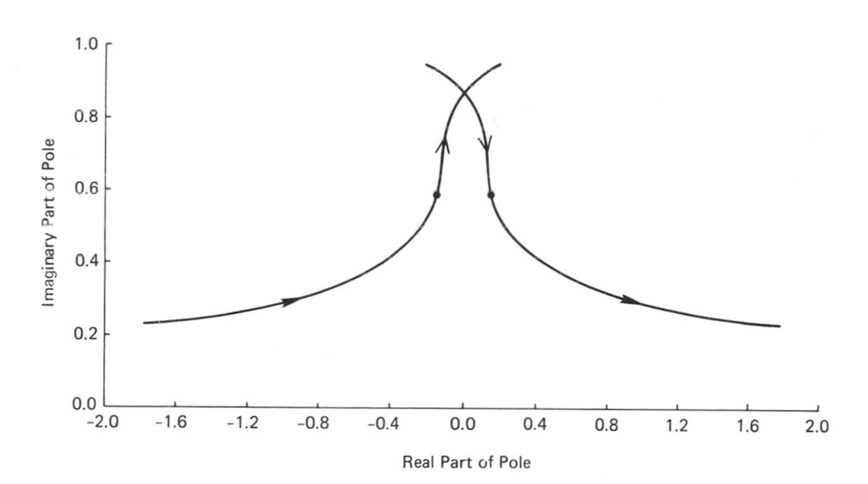

Fig. 4.5 Plots of the motions of two poles for $-0.3 \leq t \leq 0.3$. The arrows indicate the direction of motion of the poles and the dots show position of the poles at $t = 0$. The open arrow —>— denotes motion of $x_1(t)$; the solid arrow —➤— denotes motions of $x_2(t)$. (From reference [114], by permission of The Intitute of Physics, England.)

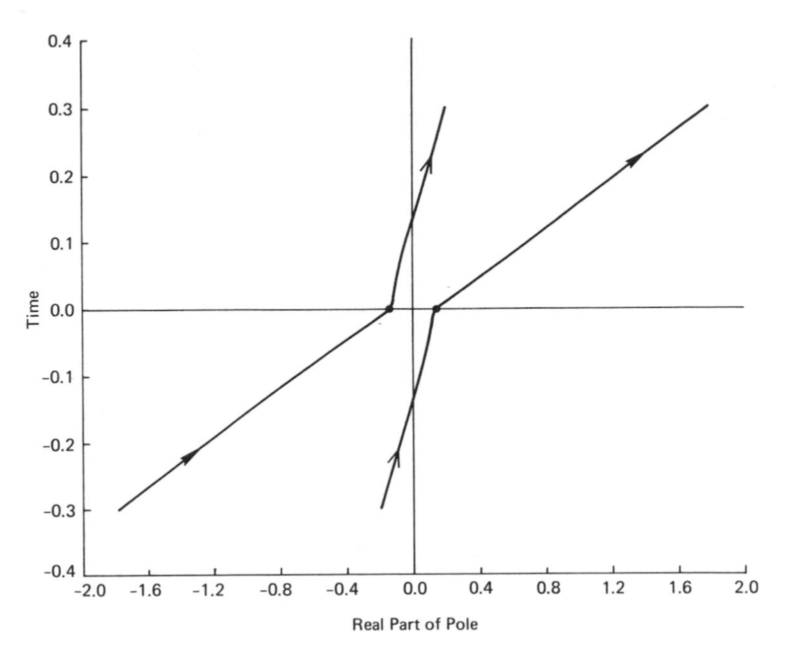

Fig. 4.6 Plot of Re $x_1(t)$ (—→—) and Re $x_2(t)$ (—➤—) as functions of time. (From reference [114], by permission of The Institute of Physics, England.)

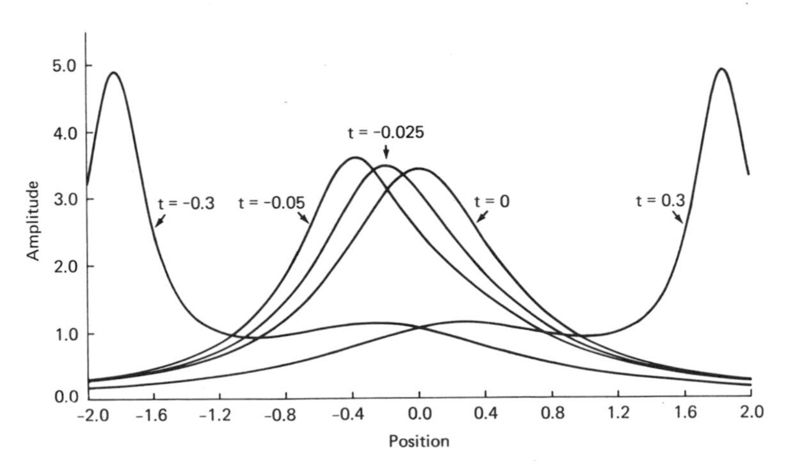

Fig. 4.7 Plot is the same as Fig. 4.4 except that $a_2 = 3 + 2\sqrt{2}$. (From reference [114], by permission of The Institute of Physics, England.)

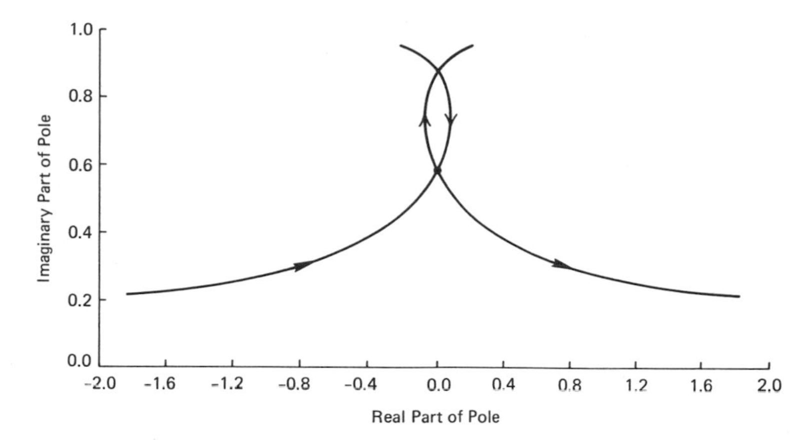

Fig. 4.8 Plot is the same as Fig. 4.5 except that $a_2 = 3 + 2\sqrt{2}$. (From reference [114], by permission of The Institute of Physics, England.)

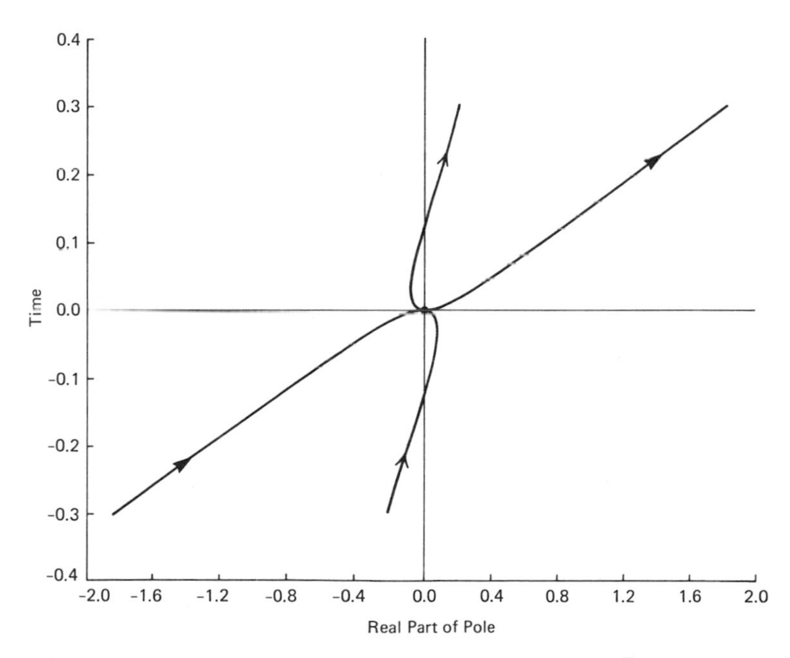

Fig. 4.9 Plot is the same as Fig. 4.6 except that $a_2 = 3 + 2\sqrt{2}$. (From reference [114], by permission of The Institute of Physics, England.)

Figures 4.7–4.9 show plots corresponding to those of Case A, with parameters given by

$$a_1 = 1.0, \tag{4.50}$$

$$a_2 = 3 + 2\sqrt{2} = 5.83, \tag{4.51}$$

$$x_{01} = x_{02} = 0, \tag{4.52}$$

$$D = 0, \tag{4.53}$$

$$l(t_c) = l(0) = 0. \tag{4.54}$$

We can see from (4.48) and (4.49) that two poles unite at $t = 0$, that is, $x_1(0) = x_2(0)$. This is marked as a dot in Fig. 4.8.

Case C. When $D < 0$, condition (4.35),

$$a(t) = -\frac{(a_1 - a_2)^2}{a_1 a_2}\frac{t}{\sqrt{-D}}\,[1 + O(t^2)], \tag{4.55}$$

$$b(t) = \sqrt{-D}[1 + O(t^2)], \tag{4.56}$$

$$x_1(t) = \left[\frac{a_1 + a_2}{2} - \frac{(a_1 - a_2)^2}{2a_1 a_2}\frac{1}{\sqrt{-D}}\right]t + i\left(\frac{a_1 + a_2}{2a_1 a_2} + \frac{\sqrt{-D}}{2}\right) + O(t^2), \tag{4.57}$$

$$x_2(t) = \left[\frac{a_1 + a_2}{2} + \frac{(a_1 - a_2)^2}{2a_1 a_2}\frac{1}{\sqrt{-D}}\right]t + i\left(\frac{a_1 + a_2}{2a_1 a_2} - \frac{\sqrt{-D}}{2}\right) + O(t^2), \tag{4.58}$$

$$l(t_c) = l(0) = \sqrt{-D}. \tag{4.59}$$

Figures 4.10–4.12 show plots corresponding to those of Case A, with parameters given by

$$a_1 = 1.0, \tag{4.60}$$

$$a_2 = 6.0, \tag{4.61}$$

$$x_{01} = x_{02} = 0, \tag{4.62}$$

$$D = -0.0544, \tag{4.63}$$

$$l(t_c) = l(0) = 0.233. \tag{4.64}$$

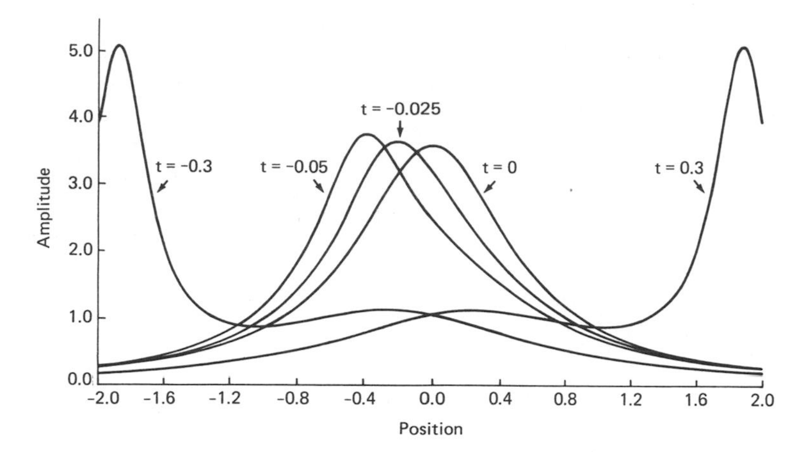

Fig. 4.10 Plot is the same as Fig. 4.4 except that $a_2 = 6.0$. (From reference [114], by permission of The Institute of Physics, England.)

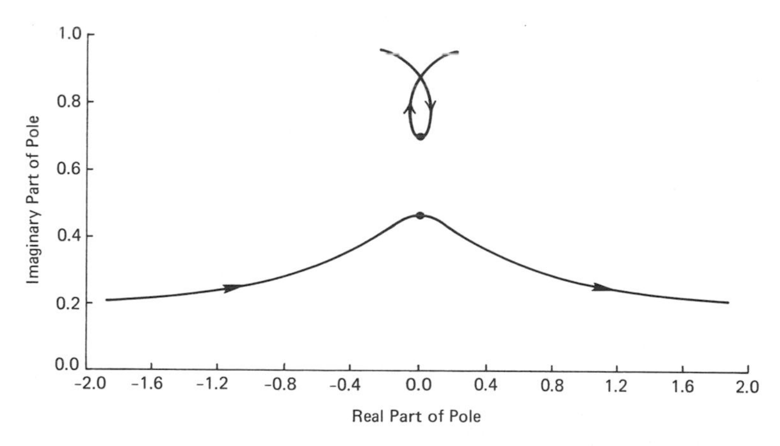

Fig. 4.11 Plot is the same as Fig. 4.5 except that $a_2 = 6.0$. (From reference [114], by permission of The Institute of Physics, England.)

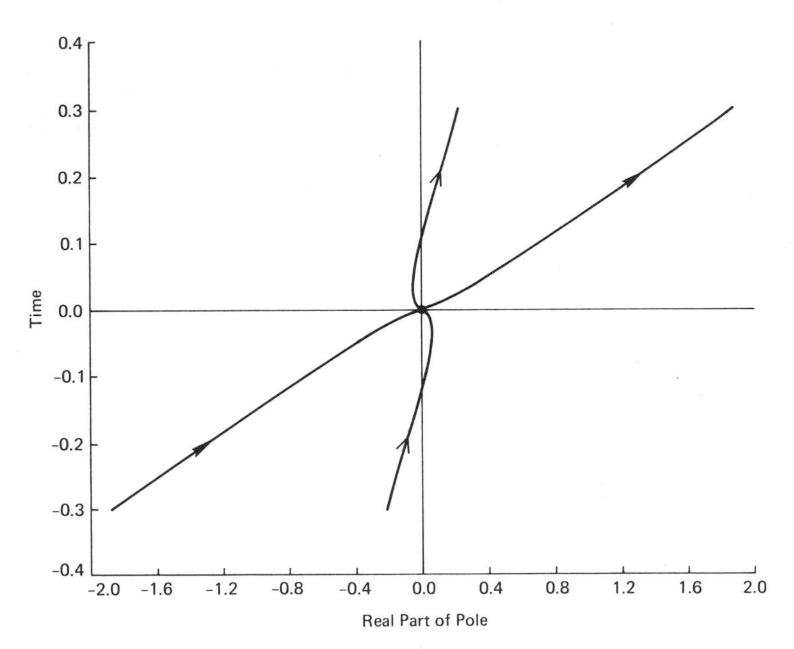

Fig. 4.12 Plot is the same as Fig. 4.6 except that $a_2 = 6.0$. (From reference [114], by permission of The Institute of Physics, England.

From these figures we find that the trajectories of two poles are continuous and have no cusps for all times, a clear difference between Cases A and C. Note also that $d\,\mathrm{Re}\,x_1(t)/dt$, which represents the velocity of the smaller soliton, becomes zero at two times (from Fig. 4.12 these times are seen to be at $t \simeq \pm 0.03$) and has a minimum value at $t = 0$ given by

$$\begin{aligned}\min d\,\mathrm{Re}\,x_1(t)/dt &= d\,\mathrm{Re}\,x_1(t)/dt\,|_{t=0} \\ &= (a_1 + a_2)/2 - (a_1 - a_2)^2/(2a_1a_2\sqrt{-D}), \quad (4.65)\end{aligned}$$

where we have used (4.57). This minimum value is -0.54 in the present example. On the other hand, $d\,\mathrm{Re}\,x_2(t)/dt$, which represents the velocity of the larger soliton, does not become zero and has a maximum value at $t = 0$ given by

$$\begin{aligned}\max d\,\mathrm{Re}\,x_2(t)/dt &= d\,\mathrm{Re}\,x_2(t)/dt\,|_{t=0} \\ &= (a_1 + a_2)/2 + (a_1 - a_2)^2/(2a_1a_2\sqrt{-D}), \quad (4.66)\end{aligned}$$

where we have used (4.58). In the present example this maximum value is 12.4.

Although the processes of the interaction of two poles are very different for the three cases A–C, the profiles of the two-soliton solution obtained are similar to one another, that is, they have only one peak at the instant of the collision (see Figs. 4.4, 4.7, and 4.10). The profile of the two-soliton solution at $t = 0$ is found from (3.9) and (3.32) with $x_{01} = x_{02} = 0$ as

$$u(x, 0) = \frac{a_1 a_2 (a_1 + a_2) x^2 + (a_1 + a_2)[(a_1 + a_2)/(a_1 - a_2)]^2}{\{a_1 a_2 x^2 - [(a_1 + a_2)/(a_1 - a_2)]^2\}^2 + (a_1 + a_2)^2 x^2}, \tag{4.67}$$

and the x derivative of $u(x, 0)$ is given by

$$\begin{aligned} u_x(x, 0) = & -2(a_1 + a_2)x\big((a_1 a_2)^3 x^4 + 2(a_1 a_2)^2[(a_1 + a_2)/(a_1 - a_2)]^2 x^2 \\ & - \{3a_1 a_2[(a_1 + a_2)/(a_1 - a_2)]^4 \\ & - (a_1 + a_2)^2[(a_1 + a_2)/(a_1 - a_2)]^2\}\big) \\ & \times \big(\{a_1 a_2 x^2 - [(a_1 + a_2)/(a_1 - a_2)]^2\}^2 + (a_1 + a_2)^2 x^2\}\big)^{-2}. \end{aligned} \tag{4.68}$$

Note also that

$$u_{xx}(0, 0) = -2a_1^3[(s - 1)^4/(s + 1)^3](s^2 - 5s + 1). \tag{4.69}$$

It follows from (4.68) and (4.69) that $u(x, 0)$ has different profiles depending on the value of a parameter $s = a_2/a_1$. For the condition

$$1 < s < (5 + \sqrt{21})/2 = 4.79, \tag{4.70}$$

the maximum value of $u(x, 0)$ is given by

$$\begin{aligned} \max u(x, 0) &= u(\pm x_0, 0) \\ &= a_1 \frac{(s - 1)^2}{s + 1} \\ &\quad \times \frac{s}{(-s^2 + 6s - 1)^{1/2}[2s^{1/2} - (-s^2 + 6s - 1)^{1/2}]}, \end{aligned} \tag{4.71}$$

where

$$x_0 = \frac{1}{a_1}\frac{1}{\sqrt{s}}\frac{s + 1}{s - 1}\left[\left(\frac{-s^2 + 6s - 1}{s}\right)^{1/2} - 1\right]^{1/2}. \tag{4.72}$$

For the condition

$$s > (5 + \sqrt{21})/2, \tag{4.73}$$

$$\max u(x, 0) = u(0, 0) = a_1(s - 1)^2/(s + 1). \tag{4.74}$$

It is interesting that for $1 < s < 3 + 2\sqrt{2}$, the positions of maximum values of $u(x, 0)$ become $\pm x_0$ [for $1 < s < (5 + \sqrt{21})/2$] and 0 [for $(5 + \sqrt{21})/2 < s < 3 + 2\sqrt{2}$]. These positions do not coincide with the positions corresponding to maximum values of two solitons, which are given by

$$\operatorname{Re} x_1(-0) = \frac{\sqrt{D}}{2} = \frac{1}{a_1} \frac{1}{\sqrt{s}} \frac{s + 1}{s - 1} \left[1 - \frac{(s - 1)^2}{4s} \right]^{1/2}, \tag{4.75}$$

$$\operatorname{Re} x_2(-0) = -\frac{\sqrt{D}}{2}. \tag{4.76}$$

It may be seen from (4.72) and (4.75) that

$$x_0 < \operatorname{Re} x_1(-0) \qquad \text{for} \quad s > 1. \tag{4.77}$$

However, the process of the interaction of two solitons for $1 < s < (5 + \sqrt{21})/2$ is not substantially different from that for $1 < s < 3 + 2\sqrt{2}$, as was shown in Case A. These circumstances are depicted in Figs. 4.13–4.15, where plots corresponding to those of Case A are shown, with parameters given by

$$a_1 = 1.0, \tag{4.78}$$

$$a_2 = 4.6, \tag{4.79}$$

$$x_{01} = x_{02} = 0, \tag{4.80}$$

$$D = 0.622, \tag{4.81}$$

$$l(t_c) = l(0) = 0.789. \tag{4.82}$$

It follows from (4.70)–(4.72), (4.75), and (4.76) that

$$x_0 = 0.215, \tag{4.83}$$

$$\max u(x, 0) = u(\pm x_0, 0) = 2.33, \tag{4.84}$$

$$\operatorname{Re} x_1(-0) = -\operatorname{Re} x_2(-0) = 0.394. \tag{4.85}$$

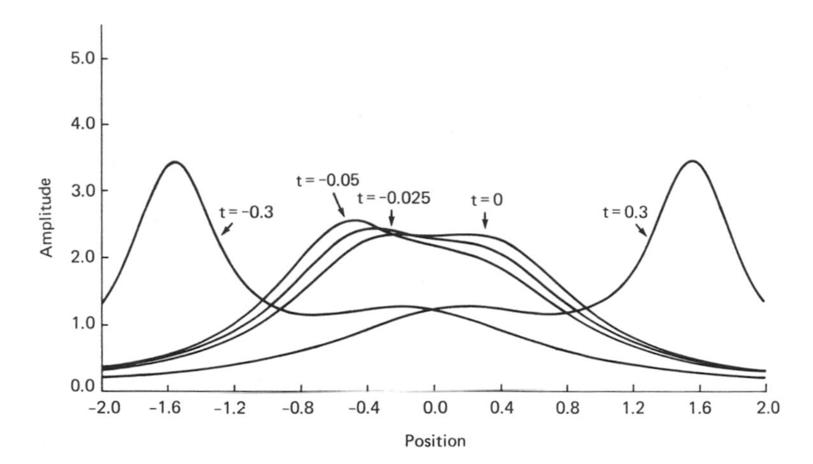

Fig. 4.13 Plot is the same as Fig. 4.4 except that $a_2 = 4.6$. (From reference [114], by permission of The Institute of Physics, England.)

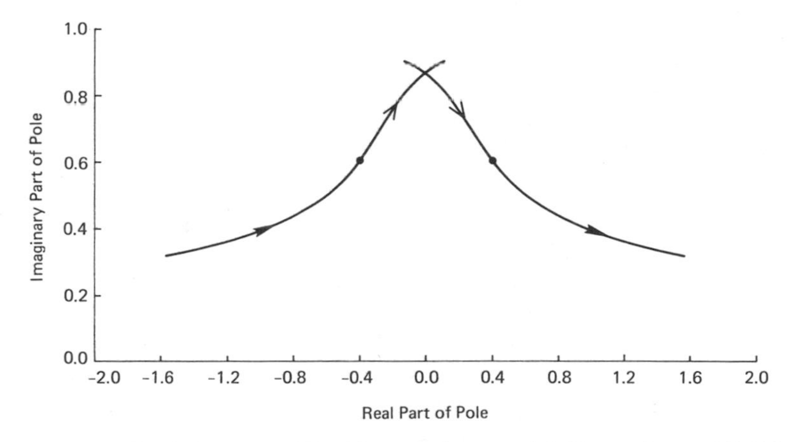

Fig. 4.14 Plot is the same as Fig. 4.5 except that $a_2 = 4.6$. (From reference [114], by permission of The Institute of Physics, England.)

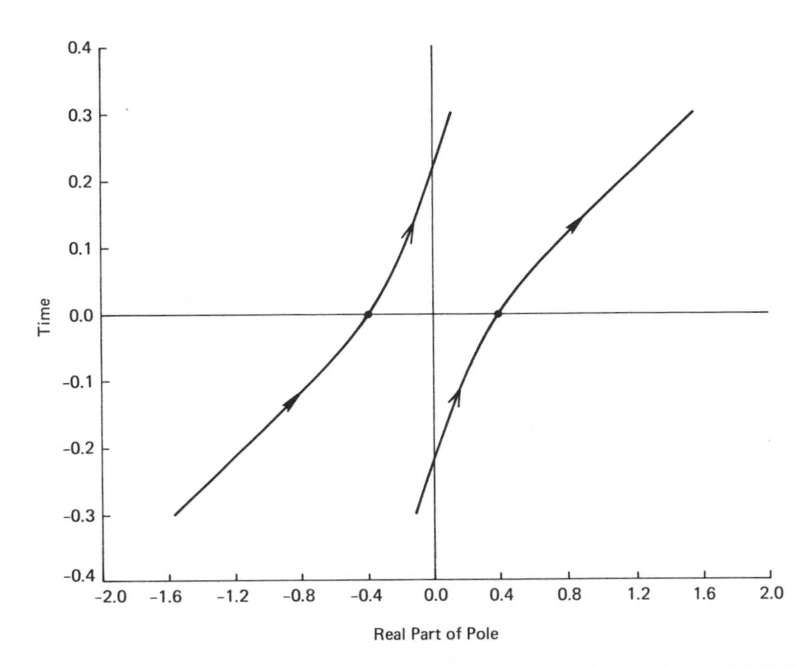

Fig. 4.15 Plot is the same as Fig. 4.6 except that $a_2 = 4.6$. (From reference [114], by permission of The Institute of Physics, England.)

The nature of the interaction of two solitons can be divided into two classes, depending on the initial amplitudes of two solitons, as follows:

(i) For $a_1 < a_2 < (3 + 2\sqrt{2})a_1$, as time goes from $-\infty$ to ∞, the amplitude of the larger soliton decreases from a_2 to a_1 while the amplitude of the smaller one increases from a_1 to a_2. The two solitons interchange their velocities at the instant of collision without passing through each other (see Figs. 4.5, 4.6, 4.14, and 4.15).

(ii) For $a_2 > (3 + 2\sqrt{2})a_1$, the larger soliton first absorbs the smaller one and then emits the smaller one backward. This situation is clear from the fact that the velocity of the smaller soliton becomes negative for some range of time (see Fig. 4.12, where this range is $-0.03 \leq t \leq 0.03$). In this case the two solitons pass through each other.

5

The Benjamin–Ono-Related Equations

In this chapter the BO-related equations are analyzed using the bilinear transformation method developed in Chapter 2. The equations treated here are the higher-order BO equations [19–21], the higher-order KdV equations [22], the finite-depth fluid equation [27] and its higher-order equations [21, 29], and the higher-order modified KdV equations [30]. A systematic method for bilinearizing these equations is developed, and the solutions for these equations are presented. Finally, the Bäcklund transformations of the higher-order KdV equations are constructed on the basis of the bilinear transformation method together with the inverse scattering transforms of the same equations [31].

5.1 Higher-Order Benjamin–Ono Equations

5.1.1 *Bilinearization*

The Lax hierarchy of the BO equation is given by

$$u_t = -(\partial/\partial x)(\delta I_n/\delta u) = -\partial K_n/\partial x, \qquad n = 3, 4, 5, \ldots, \tag{5.1}$$

with

$$K_n = \delta I_n/\delta u, \qquad n = 3, 4, 5, \ldots, \tag{5.2}$$

where I_n is the nth conserved quantity of the BO equation and $\delta/\delta u$ denotes the functional derivative defined by relation (3.212).

As shown in Section 3.2, the BO equation has an infinite number of conserved quantities [see (3.187)–(3.191)] and from these quantities the K_n defined in (5.2) are calculated as

$$K_3 = Hu_x + u^2, \tag{5.3}$$

$$K_4 = -u_{2x} + \tfrac{3}{2}Huu_x + \tfrac{3}{2}uHu_x + u^3, \tag{5.4}$$

$$\begin{aligned} K_5 = {} & -Hu_{3x} - 3uu_{2x} - 2(u_x)^2 + (Hu_x)^2 + HuHu_{2x} + 2Hu^2u_x \\ & + 2uHuu_x + 2u^2Hu_x + u^4, \end{aligned} \tag{5.5}$$

$$\begin{aligned} K_6 = {} & u_{4x} - \tfrac{5}{4}(4Hu_xu_{2x} + 2Huu_{3x} + 2u_xHu_{2x} + 2uHu_{3x} + 2u_{2x}Hu_x) \\ & + \tfrac{5}{8}[-13u(u_x)^2 - 10u^2u_{2x} + 3u(Hu_x)^2 + 2Huu_xHu_x \\ & + 2Hu^2Hu_{2x} + 4(Hu_x)(Huu_x) + 2uHuHu_{2x} + 2HuH(u_x)^2 \\ & + 2HuHuu_{2x}] + \tfrac{5}{2}(u^3Hu_x + Hu^3u_x + u^2Huu_x + uHu^2u_x) \\ & + u^5, \end{aligned} \tag{5.6}$$

where

$$u_{nx} = \partial^n u/\partial x^n, \qquad n = 0, 1, 2, \ldots. \tag{5.7}$$

For $n = 3$, (5.1) and (5.3) yield

$$u_t = -Hu_{xx} - 2uu_x, \tag{5.8}$$

which is the BO equation. In this expression the coefficient of the nonlinear term has been taken to be two instead of four.

For $n \geq 4$, the equations given by (5.1) are called the higher-order BO equations. Let us now define the rank of the polynomial K_n as the sum of the number of factors u and the number of $\partial/\partial x$, as is consistent with the scaling properties of the BO equation, where the Hilbert transform operator H does not affect the rank, since it has zero net power of x. Then the K_n defined in (5.2) is a polynomial of rank $n-1$.

In this section we shall develop a systematic method for bilinearizing the higher-order BO equations and present their N-soliton and N-periodic wave solutions. We first introduce the dependent variable transformation

$$u = i\,\partial \ln(f'/f)/\partial x, \tag{5.9}$$

with

$$f(x,t) \propto \prod_{n=1}^{N} [x - x_n(t)], \tag{5.10}$$

$$f'(x,t) \propto \prod_{n=1}^{N'} [x - x'_n(t)], \tag{5.11}$$

$$\operatorname{Im} x_n > 0, \qquad n = 1, 2, \ldots, N, \tag{5.12}$$

$$\operatorname{Im} x'_n < 0, \qquad n = 1, 2, \ldots, N', \tag{5.13}$$

where x_n and x'_n are complex functions of t and N and N' are finite or infinite positive integers. Substituting (5.9) into (5.1) and integrating by x yields

$$(iD_t f'\cdot f)/f'f = -K_n, \tag{5.14}$$

where the integration constant is assumed to be zero. To transform the nth order equation of (5.1) into the bilinear equation, we impose the following $n-3$ subsidiary conditions for u:

$$u_{\tau_j} = -\partial K_{j+2}/\partial x, \qquad j = 1, 2, \ldots, n-3, \tag{5.15}$$

where τ_j $(j = 1, 2, \ldots, n-3)$ are independent auxiliary variables. Equation (5.15) is transformed into the form

$$(iD_{\tau_j} f'\cdot f)/f'f = -K_{j+2}, \qquad j = 1, 2, \ldots, n-3, \tag{5.16}$$

by introducing (5.9) into (5.15). Under conditions (5.16), the nth order equation of (5.1) is assumed to be transformed into the bilinear equation [20, 21]

$$iD_t f'\cdot f = \alpha^{(n)} D_x^{n-1} f'\cdot f + \sum_{j=1}^{n-3} \beta_j^{(n)} D_{\tau_j} D_x^{n-j-2} f'\cdot f, \tag{5.17}$$

where $\alpha^{(n)}$ and $\beta_j^{(n)}$ $(j = 1, 2, \ldots, n-3)$ are unknown coefficients. The procedure to determine these coefficients is as follows: First, differentiating formula (App. I.3)

$$\exp[(\varepsilon D_x + \delta D_t) f' \cdot f] = \exp\left\{\sinh\left[\left(\varepsilon \frac{\partial}{\partial x} + \delta \frac{\partial}{\partial t}\right) \ln \frac{f'}{f}\right] + \cosh\left[\left(\varepsilon \frac{\partial}{\partial x} + \delta \frac{\partial}{\partial t}\right) \ln f'f\right]\right\} \quad (5.18)$$

with respect to δ and setting $\delta = 0$ gives

$$D_t \exp[(\varepsilon D_x) f' \cdot f] = \exp[(\varepsilon D_x) f' \cdot f] \frac{\partial}{\partial t}\left\{\cosh\left[\varepsilon \frac{\partial}{\partial x} \ln \frac{f'}{f}\right] + \sinh\left[\varepsilon \frac{\partial}{\partial x} \ln f'f\right]\right\}. \quad (5.19)$$

Substituting (5.9) and a relation

$$Hu = iH \frac{\partial}{\partial x} \ln \frac{f'}{f} = -\frac{\partial}{\partial x} \ln f'f \quad (5.20)$$

into (5.19) with $t = \tau_j$, using (5.16), and then equating the ε^{2n-1} terms on both sides of (5.19), we obtain

$$(D_{\tau_j} D_x^{2n-1} f' \cdot f)/f'f = i \sum_{m=0}^{n-1} {}_{2n-1}C_{2m} A_{2(n-m)-1} (K_{j+2})_{2mx} + \sum_{m=0}^{n-1} {}_{2n-1}C_{2m+1} A_{2(n-m-1)} H(K_{j+2})_{(2m+1)x}; \quad (5.21)$$

from the ε^{2n} terms we obtain

$$(D_{\tau_j} D_x^{2n} f' \cdot f)/f'f = i \sum_{m=0}^{n} {}_{2n}C_{2m} A_{2(n-m)} (K_{j+2})_{2mx} + \sum_{m=0}^{n-1} {}_{2n}C_{2m+1} A_{2(n-m)-1} H(K_{j+2})_{(2m+1)x}, \quad (5.22)$$

where A_n is defined by

$$A_n = (D_x^n f' \cdot f)/f'f, \qquad n = 1, 2, \ldots, \tag{5.23}$$

$$(K_{j+2})_{mx} = \partial^m K_{j+2}/\partial x^m, \qquad m = 0, 1, 2, \ldots, \tag{5.24}$$

and ${}_nC_m$ is the binomial coefficient

$${}_nC_m = \frac{n!}{(n-m)!m!}. \tag{5.25}$$

To express A_n in terms of $u, u_x, \ldots,$ identity (5.19) is used. Setting $t = x$ in (5.19), substituting (5.9) and (5.20) into the resultant equation, and then equating the ε^{2n-1} terms on both sides of (5.19), we obtain the recursion formula for A_{2n}:

$$\begin{aligned} A_{2n} = 1/i \sum_{m=0}^{n-1} {}_{2n-1}C_{2m} A_{2(n-m)-1} u_{2mx} \\ - \sum_{m=0}^{n-1} {}_{2n-1}C_{2m+1} A_{2(n-m-1)} Hu_{(2m+1)x}; \end{aligned} \tag{5.26}$$

and from the ε^{2n} terms, the corresponding formula for A_{2n+1}:

$$\begin{aligned} A_{2n+1} = 1/i \sum_{m=0}^{n} {}_{2n}C_{2m} A_{2(n-m)} u_{2mx} \\ - \sum_{m=0}^{n-1} {}_{2n}C_{2m+1} A_{2(n-1)-1} Hu_{(2m+1)x}. \end{aligned} \tag{5.27}$$

These recursion formulas are solved successively starting with

$$A_0 = 1, \tag{5.28}$$

$$A_1 = -iu. \tag{5.29}$$

The expressions for $A_2, \ldots, A_5$ are given as

$$A_2 = -Hu_x - u^2, \tag{5.30}$$

$$A_3 = -i(u_{2x} - 3uHu_x - u^3), \tag{5.31}$$

$$A_4 = -Hu_{3x} - 4uu_{2x} + 3(Hu_x)^2 + 6u^2Hu_x + u^4, \tag{5.32}$$

$$\begin{aligned} A_5 = -i[u_{4x} - 5uHu_{3x} - 10u_{2x}Hu_x - 10u^2u_{2x} \\ + 15u(Hu_x)^2 + 10u^3Hu_x + u^5]. \end{aligned} \tag{5.33}$$

Finally, introducing (5.14), (5.26), and (5.27) into (5.17) we obtain for odd n $(=2s+1)$

$$
\begin{aligned}
-K_{2s+1} = \alpha^{(2s+1)}A_{2s} &+ \sum_{j=1}^{s-1} \beta_{2j}^{(2s+1)}\Bigg[i \sum_{m=0}^{s-j-1} {}_{2(s-j)-1}C_{2m} \\
&\times A_{2(s-j-m)-1}(K_{2j+2})_{2mx} \\
&+ \sum_{m=0}^{s-j-1} {}_{2(s-j)-1}C_{2m+1} A_{2(s-j-1)} H(K_{2j+2})_{(2m+1)x}\Bigg] \\
&+ \sum_{j=1}^{s-1} \beta_{2j-1}^{(2s+1)}\Bigg[i \sum_{m=0}^{s-j} {}_{2(s-j)}C_{2m} A_{2(s-j-m)}(K_{2j+1})_{2mx} \\
&+ \sum_{m=0}^{s-j-1} {}_{2(s-j)}C_{2m+1} A_{2(s-j-m)-1} H(K_{2j+1})_{(2m+1)x}\Bigg]
\end{aligned}
\tag{5.34}
$$

and for even n $(=2s)$

$$
\begin{aligned}
-K_{2s} = \alpha^{(2s)}&A_{2s-1} \\
&+ \sum_{j=1}^{s-2} \beta_{2j}^{(2s)}\Bigg[i \sum_{m=0}^{s-j-1} {}_{2(s-j-1)}C_{2m} A_{2(s-m-j-1)}(K_{2j+2})_{2mx} \\
&+ \sum_{m=0}^{s-j-2} {}_{2(s-j-1)}C_{2m+1} A_{2(s-m-j-1)-1} H(K_{2j+2})_{(2m+1)x}\Bigg] \\
&+ \sum_{j=1}^{s-1} \beta_{2j-1}^{(2s)}\Bigg[i \sum_{m=0}^{s-j-1} {}_{2(s-j)-1}C_{2m} A_{2(s-j-m)-1}(K_{2j+1})_{2mx} \\
&+ \sum_{m=0}^{s-j-1} {}_{2(s-j)-1}C_{2m+1} A_{2(s-m-j-1)} H(K_{2j+1})_{(2m+1)x}\Bigg].
\end{aligned}
\tag{5.35}
$$

If we substitute (5.26), (5.27), and K_n into (5.34) and (5.35), then both sides of (5.34) and (5.35) are represented by polynomials of $u, u_x, \ldots$ with rank $2s$ and $2s-1$, respectively. Equating the coefficients of $Hu_{(2s-1)x}, \ldots, u^{2s}$ on both sides of (5.34) and those of $u_{(2s-2)x}, \ldots, u^{2s-1}$ on both sides of (5.35), we can derive the simultaneous linear algebraic equations for unknown coefficients $\alpha^{(n)}$ and $\beta_j^{(n)}$ $(j = 1, 2, \ldots, n-3)$, from which these coefficients are determined. Thus, we have established a method for bilinearizing the higher-order BO equations.

We now demonstrate the procedure for bilinearization by explicit examples. We shall take the higher-order BO equations with $n = 4, 5, 6$.

(i) $n = 4$
It follows from (5.35) with $s = 2$ that

$$-K_4 = \alpha^{(4)}A_3 + \beta_1^{(4)}[iA_1K_3 + A_0H(K_3)_x]. \tag{5.36}$$

Substitution of (5.3), (5.4), (5.29), and (5.11) into (5.36) yields

$$\begin{aligned} u_{2x} &- \tfrac{3}{2}Huu_x - \tfrac{3}{2}uHu_x - u^3 \\ &= (-i\alpha^{(4)} - \beta_1^{(4)})u_{2x} + 2\beta_1^{(4)}Huu_x \\ &\quad + (3i\alpha^{(4)} + \beta_1^{(4)})uHu_x + (i\alpha^{(4)} + \beta_1^{(4)})u^3. \end{aligned} \tag{5.37}$$

Equating the coefficients of u_{2x}, Huu_x, uHu_x, and u^3 on both sides of (5.37), we obtain

$$-i\alpha^{(4)} - \beta_1^{(4)} = 1, \tag{5.38}$$

$$2\beta_1^{(4)} = -\tfrac{3}{2}, \tag{5.39}$$

$$3i\alpha^{(4)} + \beta_1^{(4)} = -\tfrac{3}{2}, \tag{5.40}$$

$$i\alpha^{(4)} + \beta_1^{(4)} = -1, \tag{5.41}$$

from which $\alpha^{(4)}$ and $\beta_1^{(4)}$ are determined as

$$\alpha^{(4)} = i/4 \tag{5.42}$$

$$\beta_1^{(4)} = -\tfrac{3}{4}. \tag{5.43}$$

(ii) $n - 5$
It follows from (5.34) with $s = 2$ that

$$\begin{aligned} -K_5 = \alpha^{(5)}A_4 &+ \beta_1^{(5)}[iA_2K_3 + iA_0(K_3)_{2x} + 2A_1H(K_3)_x] \\ &+ \beta_2^{(5)}[iA_1K_4 + A_0H(K_4)_x]. \end{aligned} \tag{5.44}$$

Substituting (5.3)–(5.5), (5.28)–(5.30), and (5.32) into (5.44) and equating the coefficients of Hu_{3x}, uu_{2x}, $\ldots$, u^4 on both sides of (5.44), we have formally 10 equations but essentially three independent equations for three unknown coefficients $\alpha^{(5)}$, $\beta_1^{(5)}$, and $\beta_2^{(5)}$:

$$-\alpha^{(5)} + i\beta_1^{(5)} - \beta_2^{(5)} = 1, \tag{5.45}$$

$$2i\beta_1^{(5)} - \tfrac{3}{2}\beta_2^{(5)} = \tfrac{3}{2}, \tag{5.46}$$

$$3\beta_2^{(5)} = -2. \tag{5.47}$$

These equations have been obtained by comparing the coefficients of Hu_{3x}, $(u_x)^2$, and Hu^2u_x, respectively. Thus, $\alpha^{(5)}$, $\beta_1^{(5)}$, and $\beta_2^{(5)}$ are determined as

$$\alpha^{(5)} = -\tfrac{1}{12}, \tag{5.48}$$

$$\beta_1^{(5)} = -\tfrac{1}{4}i, \tag{5.49}$$

$$\beta_2^{(5)} = -\tfrac{2}{3}. \tag{5.50}$$

(iii) $n = 6$

It follows from (5.34) with $s = 3$ that

$$\begin{aligned}-K_6 = \alpha^{(6)}A_5 &+ \beta_1^{(6)}[iA_3K_3 + 3iA_1(K_3)_{2x} \\ &+ 3A_2H(K_3)_x + A_0H(K_3)_{2x}] \\ &+ \beta_2^{(6)}[iA_2K_4 + iA_0(K_4)_{2x} + 2A_1H(K_4)_x] \\ &+ \beta_3^{(6)}[iA_1K_5 + A_0H(K_5)_x].\end{aligned} \tag{5.51}$$

Substituting (5.3)–(5.6), (5.28)–(5.31), and (5.33) into (5.51) and equating the coefficients of $u_{4x}, Hu_xu_{2x}, \ldots, u^5$ on both sides of (5.51), we have formally 20 equations but essentially four independent equations for four unknown coefficients $\alpha^{(6)}$, $\beta_1^{(6)}$, $\beta_2^{(6)}$, and $\beta_3^{(6)}$:

$$-i\alpha^{(6)} - \beta_1^{(6)} - i\beta_2^{(6)} + \beta_3^{(6)} = -1, \tag{5.52}$$

$$6\beta_1^{(6)} + \tfrac{9}{2}i\beta_2^{(6)} - 5\beta_3^{(6)} = 5, \tag{5.53}$$

$$3i\beta_2^{(6)} - 3\beta_3^{(6)} = \tfrac{5}{2}, \tag{5.54}$$

$$2\beta_3^{(6)} = -\tfrac{5}{4}. \tag{5.55}$$

These equations have been obtained by comparing the coefficients of u_{4x}, Hu_xu_{2x}, u_xHu_{2x}, and Hu^2Hu_{2x}. Thus, $\alpha^{(6)}$, $\beta_1^{(6)}$, and $\beta_2^{(6)}$ are determined as

$$\alpha^{(6)} = -\tfrac{1}{96}i, \tag{5.56}$$

$$\beta_1^{(6)} = \tfrac{5}{32}, \tag{5.57}$$

$$\beta_2^{(6)} = -\tfrac{5}{24}i, \tag{5.58}$$

$$\beta_3^{(6)} = -\tfrac{5}{8}. \tag{5.59}$$

5.1.2 *Solutions*

Once the bilinearization has been performed, the solutions are constructed following the procedure developed in Chapter 2. The N-periodic wave solutions of the nth BO equation ($n = 4, 5, 6$) may be expressed compactly in the form

$$u = -\sum_{m=1}^{N} k_m + i\frac{\partial}{\partial x}\ln\frac{f^*}{f}, \tag{5.60}$$

$$f = \sum_{\mu=0,1}\exp\left[\sum_{m=1}^{N}\mu_m(i\theta_m^{(n)} + \phi_m^{(n)}) + \sum_{l<m}^{(N)}\mu_l\mu_m A_{lm}^{(n)}\right], \tag{5.61}$$

$$f' = \exp\left[\sum_{m=1}^{N}(i\theta_m^{(n)} - \phi_m^{(n)}) + \sum_{l<m}^{(N)} A_{lm}^{(n)}\right]f^*, \tag{5.62}$$

with

$$\theta_m^{(n)} = k_m\left[x - c_m^{(n)}t - \sum_{j=1}^{n-3}c_m^{(j+2)}\tau_j - \delta_m^{(n)}\right]$$
$$+ \tilde{\delta}_m^{(n)} + i\sum_{l\neq m}^{N}\tfrac{1}{2}A_{lm}^{(n)}, \qquad n = 4, 5, 6, \tag{5.63}$$

$$c_m^{(3)} = k_m \coth\phi_m^{(n)}, \qquad \phi_m^{(n)}/k_m > 0, \tag{5.64}$$

$$c_m^{(4)} = \tfrac{3}{4}[c_m^{(3)}]^2 + \tfrac{1}{4}k_m^2, \tag{5.65}$$

$$c_m^{(5)} = \tfrac{1}{2}\{c_m^{(3)}\}^3 + \tfrac{1}{2}k_m^2 c_m^{(3)}, \tag{5.66}$$

$$c_m^{(6)} = \tfrac{5}{16}[c_m^{(3)}]^4 + \tfrac{5}{8}k_m^2[c_m^{(3)}]^2 + \tfrac{1}{16}k_m^4, \qquad m = 1, 2, \ldots, N, \tag{5.67}$$

$$\exp(A_{lm}^{(n)}) = \frac{(c_l^{(n)} - c_m^{(n)})^2 - (k_l - k_m)^2}{(c_l^{(n)} - c_m^{(n)})^2 - (k_l + k_m)^2}, \tag{5.68}$$

where k_m, $\phi_m^{(n)}$, $\delta_m^{(n)}$, and $\tilde{\delta}_m^{(n)}$ ($m = 1, 2, \ldots, N$; $n = 4, 5, 6$) are arbitrary constants, $*$ denotes complex conjugate, $\sum_{\mu=0,1}$ indicates the summation over all possible combinations of $\mu_1 = 0, 1, \mu_2 = 0, 1, \ldots, \mu_N = 0, 1$, and $\sum_{l<m}^{(N)}$ means the summation over all possible combinations of N elements under the condition $l < m$.

The N-soliton solution may be obtained from (5.60)–(5.68) by setting

$$\tilde{\delta}_m^{(n)} = \pi, \qquad m = 1, 2, \ldots, N, \quad n = 4, 5, 6, \tag{5.69}$$

and taking the long-wave limit $k_m \to 0$ $(m = 1, 2, \ldots, N)$, keeping $c_m^{(3)}$ finite. The result is expressed as

$$u = i\frac{\partial}{\partial x}\ln\frac{f^*}{f}, \tag{5.70}$$

$$f = \det M, \tag{5.71}$$

where M is an $N \times N$ matrix whose elements are given by

$$M_{ml} = \begin{cases} i\left[x - \dfrac{n-1}{2^{n-2}}(c_m^{(3)})^{n-2}t - \displaystyle\sum_{j=1}^{n-3}\dfrac{j+1}{2^j}(c_j^{(3)})^j\tau_j - \delta_m^{(n)}\right] + \dfrac{1}{c_m^{(3)}} & \text{for} \quad m = l, \quad (5.72) \\ \dfrac{2}{c_m^{(3)} - c_l^{(3)}} & \text{for} \quad m \neq l, \quad (5.73) \end{cases}$$

with $c_m^{(3)}$ and $\delta_m^{(n)}$ $(n = 4, 5, 6)$ being arbitraty constants. If we set

$$\delta_m^{(n)'} = \delta_m^{(n)} + \sum_{j=1}^{n-3}\frac{j+1}{2^j}(c_j^{(3)})^j\tau_j, \tag{5.74}$$

then the functional form of the solution is the same for all n, the only difference being the velocity $(n-1)(c_m^{(3)})^{n-2}/2^{n-2}$ of the solitons.

5.2 Higher-Order Korteweg–de Vries Equations

5.2.1 Bilinearization

Consider the higher-order KdV equations

$$u_t = -(\partial/\partial x)(\delta I_n/\delta u) = -\partial K_n/\partial x, \qquad n = 3, 4, 5, \ldots, \tag{5.75}$$

where I_n is the nth conserved quantity of the KdV equation. The first few expressions of K_n are given as

$$K_3 = u_{2x} + 3u^2, \tag{5.76}$$

$$K_4 = u_{4x} + 10uu_{2x} + 5(u_x)^2 + 10u^3, \tag{5.77}$$

$$K_5 = u_{6x} + 14uu_{4x} + 28u_xu_{3x} + 21(u_{2x})^2 + 70u^2u_{2x} + 70u(u_x)^2 + 35u^4, \tag{5.78}$$

$$K_6 = u_{8x} + 18uu_{6x} + 54u_xu_{5x} + 114u_{2x}u_{4x} + 69(u_{3x})^2 + 126u^2u_{4x} + 462(u_x)^2u_{2x} + 504uu_xu_{3x} + 378u(u_{2x})^2 + 420u^3u_{2x} + 630u^2(u_x)^2 + 126u^5. \tag{5.79}$$

These expressions may be generated from the recursion formula [43, 54]

$$\partial K_{n+1}/\partial x = \partial^3 K_n/\partial x^3 + 4u\, \partial K_n/\partial x + 2u_x K_n \tag{5.80}$$

starting with $K_1 = \frac{1}{2}$. In (5.76)–(5.79), the terms K_n are normalized to make the coefficient of the highest derivative term $u_{2(n-2)x}$ equal to one since numerical factors that multiply K_n can be absorbed into the time variable in (5.75).

For $n = 3$, Eq. (5.75) with (5.76) is the KdV equation

$$u_t = -u_{3x} - 6uu_x. \tag{5.81}$$

The procedure for bilinearizing the higher-order KdV equations is the same as that for the higher-order BO equations developed in Section 5.1.

To bilinearize the nth order equation (5.75), impose $n - 3$ subsidiary conditions for u:

$$u_{\tau_j} = -\partial K_{j+2}/\partial x, \qquad j = 1, 2, \ldots, n-3, \tag{5.82}$$

where τ_j $(j = 1, 2, \ldots, n-3)$ are independent auxiliary variables. Introducing the dependent variable transformation

$$u = 2\, \partial^2 \ln f/\partial x^2, \tag{5.83}$$

into (5.82) and integrating by x yields

$$(D_{\tau_j} D_x f \cdot f)/f^2 = -K_{j+2}, \qquad j = 1, 2, \ldots, n-3. \tag{5.84}$$

Under conditions (5.82) [or (5.84)], the nth order KdV equation may be transformed into the bilinear equation [22]

$$D_t D_x f \cdot f = \alpha^{(n)} D_x^{2(n-1)} f \cdot f + \sum_{j=1}^{n-3} \beta_j^{(n)} D_{\tau_j} D_x^{2(n-j)-3} f \cdot f, \qquad n = 3, 4, 5, \ldots, \tag{5.85}$$

where $\alpha^{(n)}$ and $\beta_j^{(n)}$ $(j = 1, 2, \ldots, n-3)$ are unknown coefficients. Consider now an identity of the bilinear operators[§]

$$\exp(\varepsilon D_x) a \cdot b = [\exp(\varepsilon\, \partial/\partial x\, (a/b)][\cosh(\varepsilon D_x) b \cdot b], \tag{5.86}$$

[§] $\cosh(\varepsilon D_x) b \cdot b = \frac{1}{2}[\exp(\varepsilon D_x) + \exp(-\varepsilon D_x)] b \cdot b$.

which is derived from formulas (App. I.4) and (App. I.5). Setting $a = f_{\tau_j}$, $b = f$ and expanding the ε^{2m+1} terms, we obtain

$$\frac{1}{2}\frac{1}{(2m+1)!}D_x^{2m+1}D_{\tau_j}f\cdot f$$
$$= \sum_{s=0}^{m}\frac{1}{(2s+1)!(2m-2s)!}\left(\frac{\partial^{2s+2}\ln f}{\partial\tau_j\,\partial x^{2s+1}}\right)[D_x^{2(m-s)}f\cdot f] \quad (5.87)$$

by using formula (App. I.1.6). Here we note the relation

$$\frac{\partial^{2s+2}\ln f}{\partial\tau_j\,\partial x^{2s+1}} = \frac{1}{2}\frac{\partial^{2s-1}}{\partial x^{2s-1}}\frac{\partial u}{\partial\tau_j} = -\frac{1}{2}\frac{\partial^{2s}K_{j+2}}{\partial x^{2s}}, \quad (5.88)$$

which is derived from (5.82) and (5.83). Substitution of (5.88) into (5.87) yields

$$\frac{(D_x^{2m+1}D_{\tau_j}f\cdot f)}{f^2} = -\sum_{s=0}^{m}{}_{2m+1}C_{2s+1}\frac{\partial^{2s}K_{j+2}}{\partial x^{2s}}\frac{D_x^{2(m-s)}f\cdot f}{f^2}. \quad (5.89)$$

From (5.75) and (5.83) it also follows that

$$(D_tD_xf\cdot f)/f^2 = -K_n. \quad (5.90)$$

Substituting (5.89) and (5.90) into (5.85), we obtain the desired relation

$$K_n = -\alpha^{(n)}\frac{D_x^{2(n-1)}f\cdot f}{f^2} + \sum_{j=1}^{n-3}\beta_j^{(n)}\sum_{s=0}^{n-j-2}{}_{2n-2j-3}C_{2s+1}$$
$$\times\frac{\partial^{2s}K_{j+2}}{\partial x^{2s}}\frac{D_x^{2(n-j-s-2)}f\cdot f}{f^2}, \qquad n\geq 4, \quad (5.91)$$

which is used to determine $\alpha^{(n)}$ and $\beta_j^{(n)}$. To express $(D_x^{2m}f\cdot f)/f^2$ in terms of $u, u_x, \ldots$, formula (App. I.5) is employed together with (5.83) to give

$$\cosh(\varepsilon D_x)f\cdot f = \exp\left[2\cosh\left(\varepsilon\frac{\partial}{\partial x}\right)\ln f\right] = f^2\exp\left[\sum_{s=1}^{\infty}\frac{\varepsilon^{2s}}{(2s)!}u_{2(s-1)x}\right]. \quad (5.92)$$

Comparing the ε^{2m} terms on both sides of (5.92), we obtain a useful formula

$$\frac{D_x^{2m} f \cdot f}{f^2} = (2m)! \sum_{(m)} \prod_{l=1}^{(m)} \frac{1}{s_l!} \left[\frac{u_{2(l-1)x}}{(2l)!} \right]^{s_l}, \tag{5.93}$$

where the summation $\sum_{(m)}$ is over all sets of nonnegative integers s_l satisfying the condition

$$\sum_l l s_l = m. \tag{5.94}$$

The first five expressions for $(D_x^{2m} f \cdot f)/f^2$ are given in Appendix I [see (App. I.5.1)–(App. I.5.5)].

If we define the rank of the polynomial $u^{a_0} u_x^{a_1} \cdots u_{lx}^{a_l}$ as

$$\sum_{j=0}^{l} (1 + j/2) a_j,$$

that is, by the sum of the number of factors u and half the number of $\partial/\partial x$, then both sides of (5.91) are represented by the polynomials of u, $u_x, \ldots$ with rank $n - 1$. Equating the coefficients of u^{n-1}, $u^{n-4}(u_x)^2, \ldots$, $u_{2(n-2)x}$ on both sides of (5.91), we can derive simultaneous linear algebraic equations for unknown coefficients $\alpha^{(n)}$ and $\beta_j^{(n)}$ $(j = 1, 2, \ldots, n - 3)$. This procedure is now performed explicitly in the case of the higher-order KdV equations with $n = 4, 5, 6$.

(i) $n = 4$

In this case (5.91) becomes

$$K_4 = -\alpha^{(4)} D_x^6 f \cdot f/f^2 + \beta_1^{(4)} (3K_3 D_x^2 f \cdot f/f^2 + \partial^2 K_3/\partial x^2). \tag{5.95}$$

Substituting (5.76), (5.77), (App. I.5.1), and (App. I.5.3) into (5.95) yields

$$\begin{aligned}
u_{4x} &+ 10uu_{2x} + 5(u_x)^2 + 10u^3 \\
&= -\alpha^{(4)}(u_{4x} + 15uu_x + 15u^3) \\
&\quad + \beta_1^{(4)}[3(u_{2x} + 3u^2)u + u_{4x} + 6uu_{2x} + 6(u_x)^2] \\
&= (-\alpha^{(4)} + \beta_1^{(4)})u_{4x} + (-15\alpha^{(4)} + 9\beta_1^{(4)})uu_{2x} \\
&\quad + 6\beta_1^{(4)}(u_x)^2 + (-15\alpha^{(4)} + 9\beta_1^{(4)})u^3.
\end{aligned} \tag{5.96}$$

Comparing the coefficients of u_{4x}, uu_{2x}, $(u_x)^2$, and u^3 on both sides of (5.96), we obtain

$$-\alpha^{(4)} + \beta_1^{(4)} = 1, \tag{5.97}$$

$$-15\alpha^{(4)} + 9\beta_1^{(4)} = 10, \tag{5.98}$$

$$6\beta_1^{(4)} = 5, \tag{5.99}$$

from which $\alpha^{(4)}$ and $\beta_1^{(4)}$ are determined to be

$$\alpha^{(4)} = -\tfrac{1}{6}, \tag{5.100}$$

$$\beta_1^{(4)} = \tfrac{5}{6}. \tag{5.101}$$

(ii) $n = 5$
In this case (5.91) becomes

$$K_5 = -\alpha^{(5)} \frac{D_x^8 f\cdot f}{f^2} + \beta_1^{(5)}\left[5K_3 \frac{D_x^4 f\cdot f}{f^2} + 10\frac{\partial^2 K_3}{\partial x^2}\frac{D_x^2 f\cdot f}{f^2} + \frac{\partial^4 K_3}{\partial x^4}\right] + \beta_2^{(5)}\left(3K_4 \frac{D_x^2 f\cdot f}{f^2} + \frac{\partial^2 K_4}{\partial x^2}\right). \tag{5.102}$$

Substituting (5.76)–(5.78), (App. I.5.1), (App. I.5.2), and (App. I.5.4) we obtain, by straightforward algebra,

$$\begin{aligned} &u_{6x} + 14uu_{4x} + 28u_x u_{3x} + 21(u_{2x})^2 + 70u^2u_{2x} + 70u(u_x)^2 + 35u^4 \\ &\quad= (-\alpha^{(5)} + \beta_1^{(5)} + \beta_2^{(5)})u_{6x} + (-28\alpha^{(5)} + 16\beta_1^{(5)} + 13\beta_2^{(5)})uu_{4x} \\ &\qquad + (24\beta_1^{(5)} + 30\beta_2^{(5)})u_x u_{3x} + (-35\alpha^{(5)} + 23\beta_1^{(5)} + 20\beta_2^{(5)})(u_{2x})^2 \\ &\qquad + (-210\alpha^{(5)} + 90\beta_1^{(5)} + 60\beta_2^{(5)})u^2u_{2x} + (60\beta_1^{(5)} + 75\beta_2^{(5)})u(u_x)^2 \\ &\qquad + (-105\alpha^{(5)} + 45\beta_1^{(5)} + 30\beta_2^{(5)})u^4. \end{aligned} \tag{5.103}$$

Equating the coefficients of u_{6x}, uu_{4x}, $\ldots$, u^4 on both sides of (5.103), we obtain linear algebraic equations for $\alpha^{(5)}$, $\beta_1^{(5)}$, and $\beta_2^{(5)}$:

$$-\alpha^{(5)} + \beta_1^{(5)} + \beta_2^{(5)} = 1, \tag{5.104}$$

$$-28\alpha^{(5)} + 16\beta_1^{(5)} + 13\beta_2^{(5)} = 14, \tag{5.105}$$

$$24\beta_1^{(5)} + 30\beta_2^{(5)} = 28, \tag{5.106}$$

$$-35\alpha^{(5)} + 23\beta_1^{(5)} + 20\beta_2^{(5)} = 21, \tag{5.107}$$

$$-210\alpha^{(5)} + 90\beta_1^{(5)} + 60\beta_2^{(5)} = 70, \tag{5.108}$$

$$60\beta_1^{(5)} + 75\beta_2^{(5)} = 70, \tag{5.109}$$

$$-105\alpha^{(5)} + 45\beta_1^{(5)} + 30\beta_2^{(5)} = 35. \tag{5.110}$$

It may easily be seen that only two of seven equations are independent. Thus, the unknown coefficients $\alpha^{(5)}$, $\beta_1^{(5)}$, and $\beta_2^{(5)}$ are determined as

$$\alpha^{(5)} = \tfrac{1}{6} - \tfrac{1}{4}c_1, \tag{5.111}$$

$$\beta_1^{(5)} = \tfrac{7}{6} - \tfrac{5}{4}c_1, \tag{5.112}$$

$$\beta_2^{(5)} = c_1, \tag{5.113}$$

where c_1 is an arbitrary constant. If we do not introduce the auxiliary variable τ_2, then this constant can be assumed to be zero.

(iii) $n = 6$
In this case (5.91) becomes

$$K_6 = -\alpha^{(6)} \frac{D_x^{10} f \cdot f}{f^2} + \sum_{j=1}^{3} \beta_j^{(6)} \sum_{s=0}^{4-j} {}_{9-2j}C_{2s+1} \frac{\partial^{2s} K_{j+2}}{\partial x^{2s}} \frac{D_x^{2(4-j-s)} f \cdot f}{f^2}. \tag{5.114}$$

Substituting (5.76)–(5.79) and (App. I.5.1)–(App. I.5.5) and equating the coefficients of u_{8x}, uu_{6x}, $u_x u_{5x}$, $u_{2x}u_{4x}$, $(u_{3x})^2$, u^2u_{4x}, $(u_x)^2u_{2x}$, uu_xu_{3x}, $u(u_{2x})^2$, u^3u_{2x}, $u^2(u_x)^2$, and u^5 on both sides of (5.114), we have the 12 equations:

$$-\alpha^{(6)} + \beta_1^{(6)} + \beta_2^{(6)} + \beta_2^{(6)} + \beta_3^{(6)} = 1, \tag{5.115}$$

$$-45\alpha^{(6)} + 27\beta_1^{(6)} + 20\beta_2^{(6)} + 17\beta_3^{(6)} = 18, \tag{5.116}$$

$$36\beta_1^{(6)} + 50\beta_2^{(6)} + 56\beta_3^{(6)} = 54, \tag{5.117}$$

$$-210\alpha^{(6)} + 132\beta_1^{(6)} + 115\beta_2^{(6)} + 112\beta_3^{(6)} - 114, \tag{5.118}$$

$$60\beta_1^{(6)} + 70\beta_2^{(6)} + 70\beta_3^{(6)} = 69, \tag{5.119}$$

$$-630\alpha^{(6)} + 252\beta_1^{(6)} + 145\beta_2^{(6)} + 112\beta_3^{(6)} = 126, \tag{5.120}$$

$$210\beta_1^{(6)} + 385\beta_2^{(6)} + 490\beta_3^{(6)} = 462, \tag{5.121}$$

$$504\beta_1^{(6)} + 540\beta_2^{(6)} + 504\beta_3^{(6)} = 504, \tag{5.122}$$

$$-1575\alpha^{(6)} + 693\beta_1^{(6)} + 430\beta_2^{(6)} + 343\beta_3^{(6)} = 378, \tag{5.123}$$

$$-3150\alpha^{(6)} + 1050\beta_1^{(6)} + 500\beta_2^{(6)} + 350\beta_3^{(6)} = 420, \tag{5.124}$$

$$630\beta_1^{(6)} + 675\beta_2^{(6)} + 630\beta_3^{(6)} = 630, \tag{5.125}$$

$$-945\alpha^{(6)} + 315\beta_1^{(6)} + 150\beta_2^{(6)} + 105\beta_3^{(6)} = 126. \tag{5.126}$$

It may be seen that only three of these equations are independent. Thus, we can determine $\alpha^{(6)}$, $\beta_1^{(6)}$, $\beta_2^{(6)}$, and $\beta_3^{(6)}$ as

$$\alpha^{(6)} = -\tfrac{9}{80} + \tfrac{1}{8}c_2, \tag{5.127}$$

$$\beta_1^{(6)} = -\tfrac{11}{16} + \tfrac{7}{8}c_2, \tag{5.128}$$

$$\beta_2^{(6)} = \tfrac{63}{40} - \tfrac{7}{4}c_2, \tag{5.129}$$

$$\beta_3^{(6)} = c_2, \tag{5.130}$$

where c_2 is an arbitrary constant. If we do not introduce the auxiliary variable τ_3, then c_2 can be assumed to be zero.

5.2.2 *Solutions*

The N-soliton solutions of the higher-order KdV equations are constructed following the same procedure developed in Chapter 2 for the KdV equation itself (see Section 2.2), and these may be expressed in a compact form as

$$u = 2\,\partial^2 \ln f/\partial x^2, \tag{5.131}$$

$$f = \sum_{\mu=0,1} \exp\left[\sum_{l=1}^{N} \mu_l \theta_l^{(n)} + \sum_{l<m}^{(N)} \mu_l \mu_l A_{lm}^{(n)}\right], \tag{5.132}$$

with

$$\theta_l^{(n)} = a_l\left(x - a_l^{2(n-2)}t - \sum_{j=1}^{n-3} a_l^{2j}\tau_j - \delta_l^{(n)}\right),$$

$$l = 1, 2, \ldots, N, \quad n = 4, 5, 6, \ldots, \tag{5.133}$$

$$\exp A_{lm}^{(n)} = (a_l - a_m)^2/(a_l + a_m)^2,$$

$$l, m = 1, 2, \ldots, N, \quad n = 4, 5, 6, \ldots, \tag{5.134}$$

where a_l and $\delta_l^{(n)}$ ($l = 1, 2, \ldots, N, n = 4, 5, 6, \ldots$) are arbitrary constants. If we set

$$\delta_l^{(n)\prime} = \delta_l^{(n)} + \sum_{j=1}^{n-3} a_l^{2j}\tau_j \tag{5.135}$$

in (5.133), then the structure of the N-soliton solution (5.131) is the same for all n except the velocity $a_l^{2(n-2)}$ of solitons. The situation is similar to that of the higher-order BO equations; this is an interesting property of the Lax hierarchy of nonlinear evolution equations.

5.3 The Finite-Depth Fluid Equation and Its Higher-Order Equations

5.3.1 *Bilinearization*

Consider the finite-depth fluid equation [23–26]

$$u_t + 2uu_x + Gu_{xx} = 0, \tag{5.136}$$

with

$$Gu(x,t) = \frac{\lambda}{2} P \int_{-\infty}^{\infty} \left\{ \coth\left[\frac{\pi\lambda}{2}(x'-x)\right] - \mathrm{sgn}(x'-x) \right\} u(x',t)\,dx', \tag{5.137}$$

where

$$\mathrm{sgn}(x'-x) = \begin{cases} 1 & \text{for} \quad x < x', & (5.138\text{a}) \\ 0 & \text{for} \quad x' = x, & (5.138\text{b}) \\ -1 & \text{for} \quad x > x', & (5.138\text{c}) \end{cases}$$

and λ^{-1} is a positive parameter characterizing the depth of fluid. Physically, Eq. (5.136) describes long waves in a stratified fluid of finite depth. Mathematically, it reduces to the BO equation

$$u_t + 2uu_x + Hu_{xx} = 0 \tag{5.139}$$

in the deep-water limit $\lambda \to 0$ and to the KdV equation

$$u_{\tilde{t}} + uu_{\tilde{x}} + \tfrac{1}{3}u_{\tilde{x}\tilde{x}\tilde{x}} = 0 \tag{5.140}$$

in the shallow-water limit $\lambda \to \infty$, where new variables $\tilde{t}$ and $\tilde{x}$ defined by

$$\tilde{t} = \sqrt{\lambda}\,t, \tag{5.141}$$

$$\tilde{x} = \sqrt{\lambda}\,x \tag{5.142}$$

have been introduced in (5.140). Therefore, the finite-depth fluid equation may possess characteristics common to both the BO and KdV equations.

First, we shall bilinearize (5.136). Let us evaluate the complex integral

$$I = \frac{\lambda}{2} \int_C \coth\left[\frac{\pi\lambda}{2}(z-x)\right] \frac{f_z(z-i/\lambda)}{f(z-i/\lambda)}\,dz \tag{5.143}$$

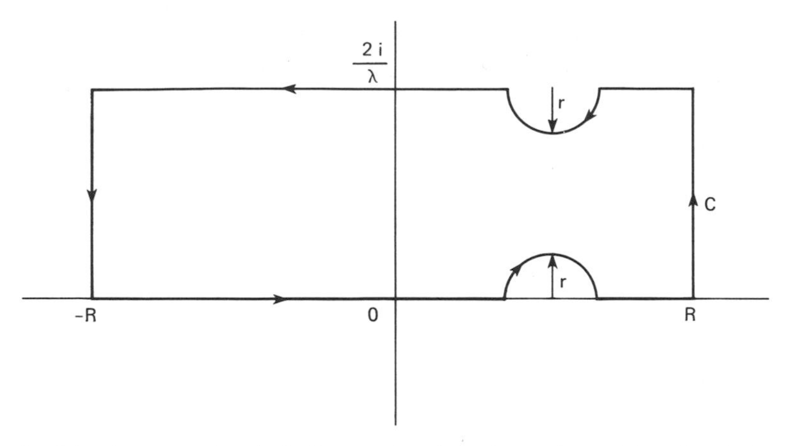

Fig. 5.1 The integral path C. See text for discussion and definition of variables.

along the integral path C shown in Fig. 5.1. It is assumed that the complex function $f(z)$ is such that $f(z - i/\lambda)$ has no zero inside C. It follows from the well-known Cauchy's residue theorem that I is modified in the limit of $R \to \infty$ and $r \to 0$ to

$$P\int_{-\infty}^{\infty} \frac{\lambda}{2} \coth\left[\frac{\pi\lambda}{2}(x' - x)\right]\left[i\frac{\partial}{\partial x}\ln\frac{f(x' - i/\lambda)}{f(x' + i/\lambda)}\right] dx'$$

$$= -\frac{\partial}{\partial x}\ln f\left(x - \frac{i}{\lambda}\right) f\left(x + \frac{i}{\lambda}\right) + c, \tag{5.144}$$

where constant c is determined by the behavior of $f(z)$ at infinity [65]. We now introduce into (5.136) the dependent variable transformation

$$u = i\frac{\partial}{\partial x}\ln\frac{f_+(x, t)}{f_-(x, t)}, \tag{5.145}$$

where

$$f_+(x, t) = f(x - i/\lambda, t), \tag{5.146}$$

$$f_-(x, t) = f(x + i/\lambda, t). \tag{5.147}$$

Then using the relation

$$Gu_x = -\frac{\partial^2}{\partial x^2}\ln f_+ f_- + \lambda u, \tag{5.148}$$

which stems from (5.144), and a formula

$$\int_{-\infty}^{\infty} \operatorname{sgn}(x' - x) u_{x'x'} \, dx' = -2 \int_{-\infty}^{\infty} \delta(x' - x) u_{x'} \, dx' = -2u_x, \quad (5.149)$$

we obtain the bilinear equation [27, 65]

$$(iD_t + i\lambda D_x - D_x^2) f_+ \cdot f_- = 0, \quad (5.150)$$

where the integration constant has been assumed to be zero. Note that, by employing formula (App. I.2), (5.150) can be rewritten in the form

$$(iD_t + i\lambda D_x - D_x^2) \exp(-i\lambda^{-1} D_x) f \cdot f = 0. \quad (5.151)$$

For real u satisfying the boundary condition $u \to 0$ as $|x| \to \infty$, f may be written in the form [27]

$$f(x, t) = \prod_{n=1}^{N} \{1 + \exp \lambda[\lambda(\operatorname{Im} x_n)x - \operatorname{Re} x_n]\}, \quad (5.152)$$

$$0 < \lambda \operatorname{Im} x_n < \pi, \qquad n = 1, 2, \ldots, N, \quad (5.153)$$

$$f_+ = f_-^*, \quad (5.154)$$

where x_n $(n = 1, 2, \ldots, N)$ are complex functions of t. The function $f(x - i/\lambda, t)$ has no zero inside C because of condition (5.153). It is now straightforward to show (5.148) directly by using (5.145), (5.152), and a formula

$$P \int_{-\infty}^{\infty} \frac{\coth[(\pi\lambda/2)(x' - x)]}{\cosh(\lambda\gamma x') + \cos(\gamma)} \, dx' = -\frac{2\lambda^{-1} \csc(\gamma) \sinh(\lambda\gamma x)}{\cosh(\lambda\gamma x) + \cos(\gamma)},$$

$$0 < \gamma < \pi. \quad (5.155)$$

The reduction of the bilinear equation (5.150) or (5.151) to the bilinearized BO and KdV equations is done as follows [27]:

(i) The BO limit $(\lambda \to 0)$
We introduce the quantities

$$\operatorname{Re} x_n = \pi \operatorname{Re} \tilde{x}_n, \quad (5.156)$$

$$\operatorname{Im} x_n = (\pi/\lambda)(1 - \lambda \operatorname{Im} \tilde{x}_n). \quad (5.157)$$

In the leading order of λ (5.152) becomes

$$f_- \simeq (-\pi\lambda)^N \prod_{n=1}^{N} (x - \tilde{x}_n) \equiv (-\pi\lambda)^N \tilde{f} \tag{5.158}$$

and (5.150) reduces to

$$(iD_t - D_x^2)\tilde{f}^* \cdot \tilde{f} = 0, \tag{5.159}$$

which is the bilinear form (3.17) of the BO equation. We note here that the condition

$$\operatorname{Im} \tilde{x}_n > 0, \qquad n = 1, 2, \ldots, N, \tag{5.160}$$

which is an assumption imposed on f in deriving (3.17), holds because of (5.153) and (5.157).

(ii) The KdV limit ($\lambda \to \infty$)
We now introduce the quantities

$$\operatorname{Re} x_n = \lambda^{-1} \operatorname{Re} \tilde{x}_n, \tag{5.161}$$

$$\operatorname{Im} x_n = \lambda^{-3/2} \operatorname{Im} \tilde{x}_n \tag{5.162}$$

together with (5.141) and (5.142). Then (5.152) becomes

$$f(x, t) = \prod_{n=1}^{N} [1 + \exp(\tilde{x} \operatorname{Im} \tilde{x}_n - \operatorname{Re} \tilde{x}_n)] \equiv \tilde{f} \tag{5.163}$$

and (5.151) reduces to

$$D_{\tilde{x}}(D_{\tilde{t}} + \tfrac{1}{3}D_{\tilde{x}}^3)\tilde{f} \cdot \tilde{f} = 0. \tag{5.164}$$

In deriving (5.164), we have used the expansion

$$\exp(-i\lambda^{-1}D_x) = \sum_{n=0}^{\infty} \frac{(-i\lambda^{-1})^n}{n!} D_x^n, \tag{5.165}$$

and formula (App. I.1.3). Equation (5.164) is the bilinearized KdV equation.

In discussing the periodic wave solution of (5.136), it is appropriate to introduce the dependent variable transformation

$$u = u_0 + i\frac{\partial}{\partial x} \ln \frac{f_+}{f_-} \tag{5.166}$$

instead of (5.145). Then (5.136) is transformed into the bilinear equation

$$[iD_t + i(\lambda + 2u_0)D_x - D_x^2 + c]f_+ \cdot f_- = 0, \tag{5.167}$$

where u_0 is a constant and c an integration constant depending generally on time. Furthermore, (5.167) can be rewritten in the form

$$\{i[D_t + (\lambda + 2u_0)D_x] \sinh(i\lambda^{-1}D_x) + (D_x^2 - c)\cosh(i\lambda^{-1}D_x)\} f \cdot f = 0 \tag{5.168}$$

using formulas (App. I.2) and (App. I.1.2). This expression will be used in Section 5.3.3.

5.3.2 *Soliton Solution*

Now we shall seek the soliton solution of (5.136) that is real and finite for all x and t and satisfies the boundary condition $u \to 0$ as $|x| \to \infty$. In this case it is convenient to employ the bilinearized equation (5.150).

For the one-soliton case, it may be confirmed by direct substitution that

$$f_- = f_+^* = 1 + e^{\lambda\gamma\theta + i\gamma} \tag{5.169}$$

satisfies (5.150). Here

$$\theta = x - at - \delta, \tag{5.170}$$

$$a = \lambda(1 - \gamma \cot \gamma), \qquad 0 < \gamma < \pi, \tag{5.171}$$

where γ and δ are constants and a is the soliton velocity. Transforming to the original variables by means of (5.145), we find

$$u = \frac{\lambda\gamma \sin \gamma}{\cosh[\lambda\gamma(x - at - \delta)] + \cos \gamma}. \tag{5.172}$$

The BO one-soliton solution is obtained from (5.172) by introducing a new constant V through the relation

$$\gamma = \pi(1 - \lambda/V) \tag{5.173}$$

and taking the deep-water limit $\lambda \to 0$ to yield

$$u = \frac{2V}{V^2(x - Vt - \delta)^2 + 1}, \tag{5.174}$$

where we have used the expansions for small λ:

$$\sin\gamma = \sin(\pi\lambda/V) \simeq \pi\lambda/V, \tag{5.175}$$

$$\cos\gamma = -\cos(\pi\lambda/V) \simeq -1 + \tfrac{1}{2}(\pi\lambda/V)^2, \tag{5.176}$$

$$a = \lambda[1 + \pi(1 - \lambda/V)\cot(\pi\lambda/V)] \simeq V, \tag{5.177}$$

$$\cosh[\lambda\gamma(x - at - \delta)] \simeq 1 + \tfrac{1}{2}(\pi\lambda)^2(x - Vt - \delta)^2. \tag{5.178}$$

On the other hand, the KdV one-soliton solution follows from (5.172) by introducing new independent variables $\tilde{t}$ and $\tilde{x}$ by (5.141) and (5.142), new constants V and δ by the relations

$$\gamma = \lambda^{-1/2}V, \tag{5.179}$$

$$\delta = \lambda^{-1/2}\tilde{\delta}, \tag{5.180}$$

and then taking the shallow-water limit $\lambda \to \infty$ to yield

$$\begin{aligned} u &= V^2 \Big/ \left\{\cosh\left[V\left(\tilde{x} - \frac{V^2}{3}\tilde{t} - \tilde{\delta}\right)\right] + 1\right\} \\ &= \frac{V^2}{2}\operatorname{sech}^2\left[\frac{V}{2}\left(\tilde{x} - \frac{V^3}{3}\tilde{t} - \tilde{\delta}\right)\right], \end{aligned} \tag{5.181}$$

with the use of the expansions for large λ:

$$\sin\gamma = \sin(\lambda^{-1/2}V) \simeq \lambda^{-1/2}V, \tag{5.182}$$

$$\cos\gamma = \cos(\lambda^{-1/2}V) \simeq 1 - V^2/2\lambda, \tag{5.183}$$

$$a = \lambda[1 - \lambda^{-1/2}V\cot(\lambda^{-1/2}V)] \simeq V^2/3. \tag{5.184}$$

The N-soliton solution is obtained similarly using the technique developed in Chapter 2, and it may be expressed in the form [27]

$$f = \sum_{\mu=0,1} \exp\left[\sum_{n=1}^{N} \mu_n(\lambda\gamma_n\theta_n + i\gamma_n) + \sum_{l<m}^{(N)} \mu_l\mu_m A_{lm}\right], \tag{5.185}$$

with

$$\theta_n = x - a_n t - \delta_n, \qquad n = 1, 2, \ldots, N, \tag{5.186}$$

$$a_n = \lambda(1 - \gamma_n\cot\gamma_n), \qquad 0 < \gamma_n < \pi, \quad n = 1, 2, \ldots, N, \tag{5.187}$$

$$e^{A_{lm}} = \frac{(a_l - a_m)^2 + \lambda^2(\gamma_l - \gamma_m)^2}{(a_l - a_m)^2 + \lambda^2(\gamma_l + \gamma_m)^2}, \tag{5.188}$$

where γ_n and δ_n ($n = 1, 2, \ldots, N$) are real constants.

The expressions of the BO and KdV N-soliton solutions are derived from (5.185) by taking the limits $\lambda \to 0$ and $\lambda \to \infty$, respectively.

(i) The BO limit ($\lambda \to 0$)
Introduce constants V_n by

$$\gamma_n = \pi(1 - \lambda/V_n), \qquad n = 1, 2, \ldots, N. \tag{5.189}$$

Following the procedure developed for the periodic wave solution of the BO equation in Section 3.1, it may be shown, in the limit of $\lambda \to 0$, that

$$f \simeq (-\pi i \lambda)^N \det M, \tag{5.190}$$

with the aid of the expansion for small λ

$$e^{A_{lm}} = 1 - 4(\pi\lambda)^2/(V_l - V_m)^2 + O(\lambda^4), \tag{5.191}$$

where M is an $N \times N$ matrix whose elements are given by

$$M_{nm} = \begin{cases} i(x - V_n t - \delta_n) + V_n^{-1} & \text{for} \quad n = m, \qquad (5.192) \\ 2/(V_n - V_m) & \text{for} \quad n \neq m. \qquad (5.193) \end{cases}$$

Except for the nonessential numerical factor $(-\pi i \lambda)^N$, expression (5.190) corresponds to the BO N-soliton solution (3.36)–(3.38).

(ii) The KdV limit ($\lambda \to \infty$)
Introduce constants V_n and $\tilde{\delta}_n$ by

$$\gamma_n = \lambda^{-1/2} V_n, \qquad n = 1, 2, \ldots, N, \tag{5.194}$$

$$\delta_n = \lambda^{-1/2} \tilde{\delta}_n, \qquad n = 1, 2, \ldots, N \tag{5.195}$$

and new variables $\tilde{t}$ and $\tilde{x}$ by (5.141) and (5.142), respectively. Then it follows from (5.185)–(5.188) that

$$f = \sum_{\mu=0,1} \exp\left[\sum_{n=1}^{N} \mu_n V_n\left(\tilde{x} - \frac{1}{3} V_n \tilde{t} - \tilde{\delta}_n\right) + \sum_{l<m}^{(N)} \mu_l \mu_m \tilde{A}_{lm}\right], \tag{5.196}$$

with

$$\tilde{A}_{lm} = (V_l - V_m)^2/(V_l + V_m)^2, \qquad l, m = 1, 2, \ldots, N, \tag{5.197}$$

which is essentially the same as the N-soliton solution of the KdV equation (2.33) and (2.34). Thus, we have demonstrated that the present N-soliton solution (5.185) with (5.186)–(5.188) reduces to both the BO and the KdV N-soliton solutions in appropriate limits.

We shall now investigate the asymptotic form of the N-soliton solution of the finite-depth fluid equation for large values of time. For this purpose, we order the parameters as

$$\gamma_1 > \gamma_2 > \cdots > \gamma_N, \tag{5.198}$$

so that in view of (5.187) a_j are ordered as

$$a_1 > a_2 > \cdots > a_N. \tag{5.199}$$

Transforming to the reference frame which moves with velocity a_n $(n = 1, 2, \ldots, N)$, we find, in the limit of $t \to -\infty$,

$$u \simeq \frac{\lambda\gamma_n \sin\gamma_n}{\cosh\{\lambda\gamma_n[x - a_n t - \delta_n + (\lambda\gamma_n)^{-1}\sum_{m=1}^{n-1} A_{nm}]\} + \cos\gamma_n}. \tag{5.200}$$

In the limit of $t \to +\infty$ we have

$$u \simeq \frac{\lambda\gamma_n \sin\gamma_n}{\cosh\{\lambda\gamma_n[x - a_n t - \delta_n + (\lambda\gamma_n)^{-1}\sum_{m=n+1}^{N} A_{nm}]\} + \cos\gamma_n}. \tag{5.201}$$

Therefore, the present solution is found to evolve asymptotically as $t \to \pm\infty$ into N localized solitons with constant velocities $a_1, a_2, \ldots, a_N$ in the original reference frame. It may be seen from (5.200) and (5.201) that, as $t \to +\infty$, the trajectory of the nth soliton is shifted by the quantity Δ_n

$$\Delta_n = \left(\sum_{m=1}^{n-1} A_{nm} - \sum_{m=n+1}^{N} A_{nm}\right)\Big/\lambda\gamma_n, \qquad n = 1, 2, \ldots, N, \tag{5.202}$$

relative to the trajectory as $t \to -\infty$.

In the BO limit ($\lambda \to 0$), (5.202) reduces, by using (5.189) and (5.191), to

$$\begin{aligned}\lim_{\lambda\to 0} \Delta_n &= \lim_{\lambda\to 0} \frac{1}{\pi\lambda}\left[\sum_{m=1}^{n-1} \frac{-4(\pi\lambda)^2}{(V_n - V_m)^2} - \sum_{m=n+1}^{N} \frac{-4(\pi\lambda)^2}{(V_n - V_m)^2} + O(\lambda^4)\right] \\ &= 0, \qquad n = 1, 2, \ldots, N.\end{aligned} \tag{5.203}$$

Thus, no phase shift results from the collision of solitons, which corresponds to the asymptotic expression (4.12).

In the KdV limit ($\lambda \to \infty$), from (5.188), (5.194), and (5.202) we obtain

$$\lim_{\lambda \to \infty} \tilde{\Delta}_n = \lim_{\lambda \to \infty} \lambda^{1/2} \Delta_n$$

$$= \frac{2}{V_n} \left[\sum_{m=1}^{n-1} \ln\left(\frac{V_m - V_n}{V_m + V_n} \right) - \sum_{m=n+1}^{N} \ln\left(\frac{V_n - V_m}{V_n + V_m} \right) \right], \quad (5.204)$$

which corresponds perfectly to formula (2.64) for the KdV equation.

5.3.3 *Periodic Wave Solution*

In this section we shall present the periodic wave solution of the finite-depth fluid equation (5.136) by employing an exact method developed in Section 2.2 [28]. It is appropriate in this case to use the bilinear equation given by (5.168), that is,

$$F(D_x, D_t) f \cdot f = 0, \quad (5.205)$$

with

$$F(D_x, D_t) = i[D_t + (\lambda + 2u_0)D_x] \sinh(i\lambda^{-1} D_x) + (D_x^2 - c) \cosh(i\lambda^{-1} D_x). \quad (5.206)$$

From the procedure developed in Section 2.2, the expression

$$f = \vartheta(\eta; \tau) = \sum_{n=-\infty}^{\infty} \exp(n\eta + n^2\tau), \qquad \tau < 0, \quad (5.207)$$

$$\eta = i(kx + \omega t + \eta_0) \quad (5.208)$$

is an exact one-periodic wave solution of (5.205), provided that the conditions

$$\tilde{F}(m) \equiv \sum_{n=-\infty}^{\infty} F[(2n - m)ik, (2n - m)i\omega] \exp\{[n^2 + (n - m)^2]\tau\}$$

$$= 0, \qquad m = 0, 1, \quad (5.209)$$

are satisfied. Here k, ω, and η_0 are wave number, frequency, and phase constant, respectively, and τ is a parameter related to the amplitude of the original wave. Introducing the notations

$$A_0 \equiv A_0(k\lambda^{-1}, \tau) = \sum_{n=-\infty}^{\infty} \cosh(2nk\lambda^{-1}) \exp(2n^2\tau), \tag{5.210}$$

$$A_1 \equiv A_1(k\lambda^{-1}, \tau) = \sum_{n=-\infty}^{\infty} \cosh[(2n-1)k\lambda^{-1}] \exp\{[n^2 + (n-1)^2]\tau\}, \tag{5.211}$$

the equations given by (5.209) may be rewritten as

$$\tilde{F}(0) = -k^2\lambda^2 A_{0,kk} + [\omega + (2u_0 + \lambda)]kA_{0,k} + cA_0 = 0 \tag{5.212}$$

$$\tilde{F}(1) = -k^2\lambda^2 A_{1,kk} + [\omega + (2u_0 + \lambda)]kA_{1,k} + cA_1 = 0, \tag{5.213}$$

where

$$A_{j,k} = \partial A_j/\partial k, \qquad j = 0, 1 \tag{5.214}$$

$$A_{j,kk} = \partial^2 A_j/\partial k^2, \qquad j = 0, 1. \tag{5.215}$$

It follows from (5.212) and (5.213) that

$$\begin{aligned}\omega &= k^2\lambda \frac{A_{0,kk}A_1 - A_0 A_{1,kk}}{A_{0,k}A_1 - A_0 A_{1,k}} - (2u_0 + \lambda)k \\ &= k^2\lambda \frac{\partial}{\partial k} \ln D_k A_0 \cdot A_1 - (2u_0 + \lambda)k,\end{aligned} \tag{5.216}$$

where D_k denotes a bilinear operator with a variable k. Using (5.210) and (5.211), we obtain the series expansion of $D_k A_0 \cdot A_1$ as

$$\begin{aligned}D_k A_0 \cdot A_1 &= \sum_{n,n'=-\infty}^{\infty} \{D_k \cosh(2nk\lambda^{-1}) \cdot \cosh[(2n'-1)k\lambda]\} \\ &\quad \times \exp\{[2n^2 + n'^2 + (n'-1)^2]\tau\} \\ &= \frac{1}{2\lambda} \sum_{n,n'=-\infty}^{\infty} \{(2n + 2n' - 1)\sinh[(2n - 2n' + 1)k\lambda^{-1}] \\ &\quad + (2n - 2n' + 1)\sinh[(2n + 2n' - 1)k\lambda^{-1}]\} \\ &\quad \times \exp\{[2n^2 + n'^2 + (n'-1)^2]\tau\} \\ &= \lambda^{-1}\{-2\sinh(k\lambda^{-1})e^{\tau} \\ &\quad + 2[3\sinh(k\lambda^{-1}) + \sinh(3k\lambda^{-1})]e^{3\tau} \\ &\quad - 6\sinh(3k\lambda^{-1})e^{5\tau} + \cdots\}.\end{aligned} \tag{5.217}$$

Substituting (5.217) into (5.216) yields

$$\omega = k^2 \frac{-2\cosh(k\lambda^{-1})e^{\tau} + 6[\cosh(k\lambda^{-1}) + \cosh(3k\lambda^{-1})]e^{3\tau} - 18\cosh(3k\lambda^{-1})e^{5\tau} + \cdots}{-2\sinh(k\lambda^{-1})e^{\tau} + 2[3\sinh(k\lambda^{-1}) + \sinh(3k\lambda^{-1})]e^{3\tau} - 6\sinh(3k\lambda^{-1})e^{5\tau} + \cdots} - (2u_0 + \lambda)k. \tag{5.218}$$

Expression (5.207), with (5.208) and (5.218), constitutes the one-periodic wave solution of (5.205).

Now consider the two limiting cases of $\lambda \to 0$ and $\lambda \to \infty$.

(i) The BO limit ($\lambda \to 0$)

We now introduce a new amplitude parameter by

$$\tau = \tau' - k\lambda^{-1}, \qquad \tau' < 0. \tag{5.219}$$

Substituting (5.219) into (5.207) and (5.218), we find, in the limit of $\lambda \to 0$,

$$\lim_{\lambda\to 0} f_{\mp} = \lim_{\lambda\to 0} f(x \pm i\lambda^{-1}, t) = 1 + \exp(\mp\eta + \tau'), \tag{5.220}$$

$$\lim_{\lambda\to 0} \omega = k^2 \frac{3e^{\tau'} - e^{(-\tau')}}{e^{\tau'} - e^{(-\tau')}} - 2u_0 k = -k^2 \coth\tau', \qquad \text{when} \quad u_0 = k, \tag{5.221}$$

which, in (5.166), produce the one-periodic wave solution (3.57) of the BO equation written in the original variable.

(ii) The KdV limit ($\lambda \to \infty$)

Rescaling wave number and frequency by

$$k = \tilde{k}\lambda^{1/2} \tag{5.222}$$

and

$$\omega = \tilde{\omega}\lambda^{1/2} \tag{5.223}$$

and space and time coordinates by (5.142) and (5.141), respectively, and substituting these into (5.207) and (5.218), we find in the limit of $\lambda \to \infty$

$$\tilde{f} = \lim_{\lambda\to\infty} f = \sum_{n=-\infty}^{\infty} \exp(n\tilde{\eta} + n^2\tau), \tag{5.224}$$

$$\tilde{\eta} = i(\tilde{k}\tilde{x} + \tilde{\omega}\tilde{t} + \eta_0), \tag{5.225}$$

$$\omega = -2u_0\tilde{k} + \frac{1}{3}\frac{1 - 30q^2 + 81q^4 + \cdots}{1 - 6q^2 + 9q^4 + \cdots}\tilde{k}^3, \qquad q \equiv e^{\tau}, \tag{5.226}$$

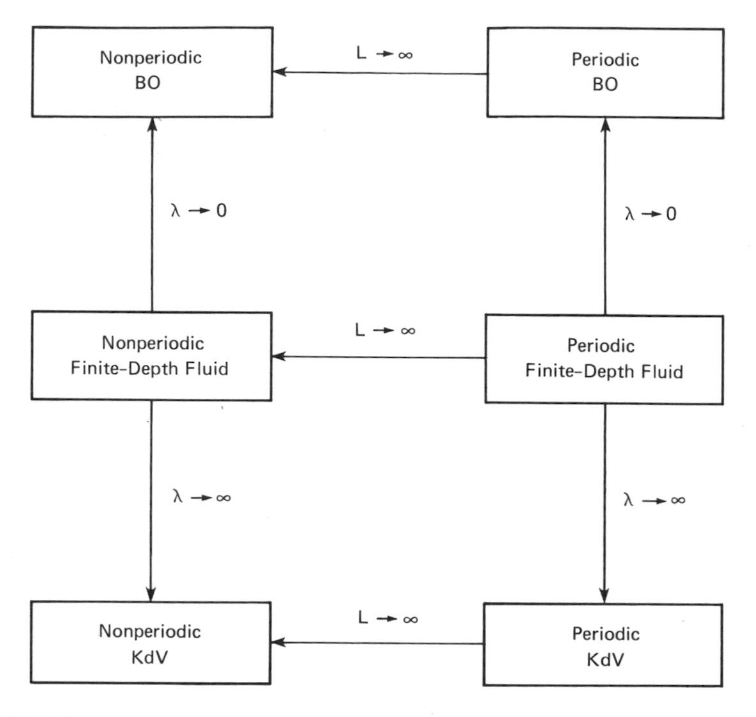

Fig. 5.2 Interrelationship among six types of solutions. Here L denotes period and λ the (depth of fluid)$^{-1}$.

which is essentially the one-periodic wave solution of the KdV equation presented in Section 2.2. This argument may be extended to the two-periodic wave solution, which is detailed in [28]. The interrelationship among solutions of the KdV, the finite-depth fluid, and the BO equations, for both periodic and nonperiodic cases, is schematically represented in Fig. 5.2.

5.3.4 Higher-Order Equations

The Lax hierarchy of the finite-depth fluid equation (5.136) is expressed as

$$u_t = -(\partial/\partial x)(\delta I_n/\delta u) = -\partial K_n/\partial x, \qquad n = 3, 4, 5, \ldots, \quad (5.227)$$

where I_n is the nth conserved quantity of the finite-depth fluid equation, the first few of which are given as [115]

$$I_1 = \int_{-\infty}^{\infty} u \, dx, \tag{5.228}$$

$$I_2 = \int_{-\infty}^{\infty} \tfrac{1}{2}u^2 \, dx, \tag{5.229}$$

$$I_3 = \int_{-\infty}^{\infty} (\tfrac{1}{3}u^3 + \tfrac{1}{2}uGu_x) \, dx, \tag{5.230}$$

$$I_4 = \int_{-\infty}^{\infty} [\tfrac{1}{4}u^4 + \tfrac{3}{4}u^2Gu_x + \tfrac{1}{8}(u_x)^2 + \tfrac{3}{8}(Gu_x)^2 + \tfrac{3}{8}\lambda uGu_x] \, dx, \tag{5.231}$$

$$\begin{aligned} I_5 = \int_{-\infty}^{\infty} \{ & \tfrac{1}{5}u^5 + \tfrac{2}{3}u^3Gu_x - \tfrac{1}{2}uu_x Gu^2 + \tfrac{1}{2}u(u_x)^2 + \tfrac{1}{2}u(Gu_x)^2 \\ & + \tfrac{1}{2}u^2G^2u_{xx} - \tfrac{1}{4}uGu_{xxx} + \tfrac{1}{4}uG^3u_{xxx} \\ & + \lambda[\tfrac{3}{4}u^2Gu_x + \tfrac{5}{24}(u_x)^2 + \tfrac{5}{8}(Gu_x)^2] + \tfrac{5}{8}\lambda^2 uGu_x\} \, dx. \end{aligned} \tag{5.232}$$

For $n = 3$, (5.227) and (5.230) yield the finite-depth fluid equation (5.136). For $n \geq 4$, Eqs. (5.227) are called higher-order finite-depth fluid equations. For $n = 4$, we find from (5.227) and (5.231) the first higher-order equation as

$$u_t = -(\partial/\partial x)(u^3 + \tfrac{3}{2}uGu_x + \tfrac{3}{2}Guu_x - \tfrac{1}{4}u_{xx} + \tfrac{3}{4}G^2u_{xx} + \tfrac{3}{4}\lambda Gu_x). \tag{5.233}$$

In this section we shall outline a method for bilinearizing (5.227). As an example, the first higher-order equation will be bilinearized and its N-soliton solution will be presented.

The procedure for the bilinearization is the same as that for the higher-order BO and KdV equations. First, impose $n - 3$ subsidiary conditions for u:

$$u_{\tau_j} = -\partial K_{j+2}/\partial x, \qquad j = 1, 2, \ldots, n-3, \tag{5.234}$$

with τ_j ($j = 1, 2, \ldots, n-3$) being independent auxiliary variables. Equations (5.234) may be transformed into the form

$$(iD_{\tau_j} f_+ \cdot f_-)/f_+ f_- = -K_{j+2}, \qquad j = 1, 2, \ldots, n-3, \tag{5.235}$$

by the dependent variable transformation (5.145). Under conditions (5.235), the nth-order equation of (5.227) may be bilinearized as

$$iD_t f_+ \cdot f_- = \sum_{j=0}^{n-2} \alpha_j^{(n)} \lambda^j D_x^{n-j-1} f_+ \cdot f_- + \sum_{j=1}^{n-3} \beta_j^{(n)} \sum_{s=0}^{n-j-3} \lambda^s \gamma_s^{(n)} D_{\tau_j} D_x^{n-j-s-2} f_+ \cdot f_-, \qquad \gamma_0^{(n)} \equiv 1, \tag{5.236}$$

where $\alpha_j^{(n)}$, $\beta_j^{(n)}$, and $\gamma_s^{(n)}$ are unknown constants. The procedure to determine these unknown constants, using (5.19) and (5.148), is the same as that for the higher-order BO equations. Instead of treating the general case, we shall illustrate the procedure in the case of $n = 4$.

From the ε term of (5.19) with $f' = f_+, f = f_-$, and $t = \tau_1$, we have

$$\frac{D_{\tau_1} D_x f_+ \cdot f_-}{f_+ f_-} = \frac{\partial^2}{\partial \tau_1 \, \partial x} \ln f_+ f_- + \frac{D_x f_+ \cdot f_-}{f_+ f_-} \frac{\partial}{\partial \tau_1} \ln \frac{f_+}{f_-}. \tag{5.237}$$

Substituting (5.148), (5.234), and (5.235) into (5.237) yields

$$\begin{aligned}\frac{D_{\tau_1} D_x f_+ \cdot f_-}{f_+ f_-} &= -Gu_{\tau_1} + \lambda \int_{-\infty}^{x} u_{\tau_1} \, dx' + uK_3 = GK_{3,x} - \lambda K_3 + uK_3 \\ &= u^3 + G^2 u_{xx} + 2Guu_x + uGu_x - \lambda u^2 - \lambda Gu_x. \end{aligned} \tag{5.238}$$

It follows from the ε^0, ε^1, and ε^2 terms of (5.19), with $f' = f_+, f = f_-$, and $t = x$, that

$$\frac{D_x f_+ \cdot f_-}{f_+ f_-} = \frac{\partial}{\partial x} \ln \frac{f_+}{f_-}, \tag{5.239}$$

$$\begin{aligned}\frac{D_x^2 f_+ \cdot f_-}{f_+ f_-} &= \frac{D_x f_+ \cdot f_-}{f_+ f_-} \frac{\partial}{\partial x} \ln \frac{f_+}{f_-} + \frac{\partial^2}{\partial x^2} \ln f_+ f_- \\ &= \left(\frac{\partial}{\partial x} \ln \frac{f_+}{f_-} \right)^2 + \frac{\partial^2}{\partial x^2} \ln f_+ f_-, \end{aligned} \tag{5.240}$$

$$\begin{aligned}\frac{D_x^3 f_+ \cdot f_-}{f_+ f_-} &= \frac{\partial^3}{\partial x^3} \ln \frac{f_+}{f_-} + 2 \frac{D_x f_+ \cdot f_-}{f_+ f_-} \frac{\partial^2}{\partial x^2} \ln f_+ f_- \\ &\quad + \frac{D_x^2 f_+ \cdot f_-}{f_+ f_-} \frac{\partial}{\partial x} \ln \frac{f_+}{f_-} \\ &= \frac{\partial^3}{\partial x^3} \ln \frac{f_+}{f_-} + 3 \frac{\partial}{\partial x} \ln \frac{f_+}{f_-} \frac{\partial^2}{\partial x^2} \ln f_+ f_- + \left(\frac{\partial}{\partial x} \ln \frac{f_+}{f_-} \right)^3. \end{aligned} \tag{5.241}$$

Substituting (5.145) and (5.148) into these expressions, we obtain

$$\frac{D_x f_+ \cdot f_-}{f_+ f_-} = -iu, \tag{5.242}$$

$$\frac{D_x^2 f_+ \cdot f_-}{f_+ f_-} = -Gu_x + \lambda u - u^2, \tag{5.243}$$

$$\frac{D_x^3 f_+ \cdot f_-}{f_+ f_-} = -i(u_{xx} - 3uGu_x + 3\lambda u^2 - u^3). \tag{5.244}$$

The expressions for $D_x^n f_+ \cdot f_-/f_+ f_-$ ($n = 1, 2, 3$) are obtained from (5.29)–(5.31) by the formal replacement

$$Hu_x \to Gu_x - \lambda u. \tag{5.245}$$

Noting the relation

$$i\frac{D_t f_+ \cdot f_-}{f_+ f_-} = -K_n, \qquad n = 3, 4, \ldots, \tag{5.246}$$

which is derived from (5.227) and (5.145), Eq. (5.236) with $n = 4$ becomes

$$-K_4 = \alpha_0^{(4)} \frac{D_x^3 f_+ \cdot f_-}{f_+ f_-} + \lambda\alpha_1^{(4)} \frac{D_x^2 f_+ \cdot f_-}{f_+ f_-} + \lambda^2\alpha_2^{(4)} \frac{D_x f_+ \cdot f_-}{f_+ f_-} + \beta_1^{(4)} \frac{D_{\tau_1} D_x f_+ \cdot f_-}{f_+ f_-}. \tag{5.247}$$

Substituting (5.238), (5.242)–(5.244), and

$$K_4 = u^3 + \tfrac{3}{2}uGu_x + \tfrac{3}{2}Guu_x - \tfrac{1}{4}u_{xx} + \tfrac{3}{4}G^2u_{xx} + \tfrac{3}{4}\lambda Gu_x \tag{5.248}$$

into (5.247), we have

$$\begin{aligned}
-(u^3 &+ \tfrac{3}{2}uGu_x + \tfrac{3}{2}Guu_x - \tfrac{1}{4}u_{xx} + \tfrac{3}{4}G^2u_{xx} + \tfrac{3}{4}\lambda Gu_x) \\
&= -i\alpha_0^{(4)}(u_{xx} - 3uGu_x + 3\lambda u^2 - u^3) + \lambda\alpha_1^{(4)}(-Gu_x + \lambda u - u^2) \\
&\quad + \lambda^2\alpha_2^{(4)}(-iu) \\
&\quad + \beta_1^{(4)}(u^3 + G^2u_{xx} + 2Guu_x + uGu_x - \lambda u^2 - \lambda Gu_x) \\
&= (i\alpha_0^{(4)} + \beta_1^{(4)})u^3 + (3i\alpha_0^{(4)} + \beta_1^{(4)})uGu_x + 2\beta_1^{(4)}Guu_x - i\alpha_0^{(4)}u_{xx} \\
&\quad + \beta_1^{(4)}G^2u_{xx} + \lambda(-\alpha_1^{(4)} - \beta_1^{(4)})Gu_x \\
&\quad + \lambda(-3i\alpha_0^{(4)} - \alpha_1^{(4)} - \beta_1^{(4)})u^2 + \lambda^2(\alpha_1^{(4)} - i\alpha_2^{(4)})u.
\end{aligned} \tag{5.249}$$

Comparing the coefficients of $u^3, uGu_x, \ldots, u$ on both sides of (5.249), we obtain the linear algebraic equations for $\alpha_j^{(4)}$ $(j = 0, 1, 2)$ and $\beta_1^{(4)}$ as

$$i\alpha_0^{(4)} + \beta_1^{(4)} = -1, \tag{5.250}$$

$$3i\alpha_0^{(4)} + \beta_1^{(4)} = -\tfrac{3}{2}, \tag{5.251}$$

$$2\beta_1^{(4)} = -\tfrac{3}{2}, \tag{5.252}$$

$$-i\alpha_0^{(4)} = \tfrac{1}{4}, \tag{5.253}$$

$$\beta_1^{(4)} = -\tfrac{3}{4}, \tag{5.254}$$

$$-\alpha_1^{(4)} - \beta_1^{(4)} = -\tfrac{3}{4}, \tag{5.255}$$

$$-3i\alpha_0^{(4)} - \alpha_1^{(4)} - \beta_1^{(4)} = 0, \tag{5.256}$$

$$\alpha_1^{(4)} - i\alpha_2^{(4)} = 0, \tag{5.257}$$

from which $\alpha_j^{(4)}$ and $\beta_1^{(4)}$ are determined to be

$$\alpha_0^{(4)} = \tfrac{1}{4}i \tag{5.258}$$

$$\alpha_1^{(4)} = \tfrac{3}{2}, \tag{5.259}$$

$$\alpha_2^{(4)} = -\tfrac{3}{2}i, \tag{5.260}$$

$$\beta_1^{(4)} = -\tfrac{3}{4}. \tag{5.261}$$

Thus, the first higher-order finite-depth fluid equation has been bilinearized as

$$iD_t f_+ \cdot f_- = (\tfrac{1}{4}iD_x^3 + \tfrac{3}{2}\lambda D_x^2 - \tfrac{3}{2}i\lambda^2 D_x - \tfrac{3}{4}D_{\tau_1}D_x)f_+ \cdot f_- \tag{5.262}$$

with a subsidiary condition

$$iD_{\tau_1} f_+ \cdot f_- = (D_x^2 - i\lambda D_x)f_+ \cdot f_- . \tag{5.263}$$

Using (5.263), (5.262) can be rewritten as [29]

$$iD_t f_+ \cdot f_- = [\tfrac{1}{4}iD_x^3 - \tfrac{3}{4}D_{\tau_1}(D_x - 2i\lambda)]f_+ \cdot f_- . \tag{5.264}$$

The higher-order equations for $n \geq 5$ can be bilinearized similarly, but we shall not discuss them here. Instead, we shall show that the nth-order equation (5.236) reduces to the nth-order BO and KdV equations.

(i) The BO limit ($\lambda \to 0$)
In the deep-water limit $\lambda \to 0$, (5.236) reduces to

$$iD_t \tilde{f}^* \cdot \tilde{f} = \alpha_0^{(n)} D_x^{n-1} \tilde{f}^* \cdot \tilde{f} + \sum_{j=0}^{n-3} \beta_j^{(n)} D_{\tau_j} D_x^{n-j-2} \tilde{f}^* \cdot \tilde{f}, \tag{5.265}$$

where $\tilde{f}$ is defined by (5.158). Obviously, (5.265) corresponds to the bilinear form of the nth order BO equation (5.17).

(ii) The KdV limit ($\lambda \to \infty$)

In the shallow-water limit $\lambda \to \infty$, it is appropriate to introduce scalings (5.141) and (5.142) together with

$$\tilde{\tau}_j = \lambda^{1/2}\tau_j, \qquad j = 1, 2, \ldots, n-3. \tag{5.266}$$

It follows by noting formulas (App. I.1.3) and (App. I.2) that

$$\begin{aligned} iD_t f_+ \cdot f_- &= iD_t \exp(-i\lambda^{-1}D_x)f \cdot f = i\lambda^{1/2}D_{\tilde{t}} \exp(-i\lambda^{-1/2}D_{\tilde{x}})f \cdot f \\ &= \sum_{s=0}^{\infty} \frac{(-1)^s\lambda^{-s}}{(2s+1)!} D_{\tilde{t}} D_{\tilde{x}}^{2s+1} f \cdot f, \end{aligned} \tag{5.267}$$

$$\begin{aligned} &\sum_{j=0}^{n-2} \alpha_j^{(n)} \lambda^j D_x^{n-j-1} f_+ \cdot f_- \\ &\quad = \sum_{j=0}^{n-2} \alpha_j^{(n)} \lambda^{(n-j-1)/2} D_{\tilde{x}}^{n-j-1} \exp(-i\lambda^{-1/2}D_{\tilde{x}}) f \cdot f \\ &\quad = \sum_{j=0}^{n-2} \alpha_j^{(n)} \sum_{s=0}^{\infty} \frac{(-i)^s}{s!} \lambda^{(n-j-s-1)/2} D_{\tilde{x}}^{n-j+s-1} f \cdot f, \end{aligned} \tag{5.268}$$

$$\begin{aligned} &\sum_{j=1}^{n-3} \beta_j^{(n)} \sum_{s=0}^{n-j-3} \lambda^s \gamma_s^{(n)} D_{\tau_j} D_x^{n-j-s-2} f_+ \cdot f_- \\ &\quad = \sum_{j=1}^{n-3} \beta_j^{(n)} \sum_{s=0}^{n-j-3} \lambda^{(n-j+s-1)/2} \gamma_s^{(n)} D_{\tilde{\tau}_j} D_{\tilde{x}}^{n-j-s-2} \exp(-i\lambda^{-1/2}D_{\tilde{x}}) f \cdot f \\ &\quad = \sum_{j=1}^{n-3} \beta_j^{(n)} \sum_{s=0}^{n-j-3} \gamma_s^{(n)} \sum_{m=0}^{\infty} \frac{(-i)^m}{m!} \lambda^{(n-j+s-m-1)/2} D_{\tilde{\tau}_j} D_{\tilde{x}}^{n-j-s+m-2} f \cdot f. \end{aligned} \tag{5.269}$$

Substituting (5.267)–(5.269) and comparing the λ^0 term on both sides of (5.236), we obtain, in the limit of $\lambda \to \infty$,

$$\begin{aligned} D_{\tilde{t}} D_{\tilde{x}} f \cdot f &= \left[\sum_{j=0}^{n-2} \alpha_j^{(n)} \frac{(-i)^{n-j-1}}{(n-j-1)!}\right] D_{\tilde{x}}^{2(n-1)} f \cdot f \\ &\quad + \sum_{j=1}^{n-3} \left[\beta_j^{(n)} \sum_{s=0}^{n-j-3} \frac{(-i)^{n-j+s-1}}{(n-j+s-1)!}\right] D_{\tilde{\tau}_j} D_{\tilde{x}}^{2(n-j)-3} f \cdot f. \end{aligned} \tag{5.270}$$

If we set

$$\tilde{\alpha}^{(n)} = \sum_{j=0}^{n-2} \frac{(-i)^{n-j-1}}{(n-j-1)!} \alpha_j^{(n)}, \tag{5.271}$$

$$\tilde{\beta}_j^{(n)} = \beta_j^{(n)} \sum_{s=0}^{n-j-3} \frac{(-i)^{n-j+s-1}}{(n-j+s-1)!}, \qquad j = 1, 2, \ldots, n-3, \tag{5.272}$$

Eq. (5.270) has the same form as the bilinear form of the nth-order KdV equation (5.85). Once the bilinear equation is obtained, the solutions are constructed following the procedure developed in Chapter 2. For the first higher-order finite-depth fluid equation in the bilinear form, (5.262) with (5.263), the N-soliton solution may be expressed in the form [29]

$$f = \sum_{\mu=0,1} \exp\left[\sum_{n=1}^{N} \mu_n(\lambda\gamma_n\theta_n + i\gamma_n) + \sum_{l<m}^{(N)} \mu_l\mu_m A_{lm} \right], \tag{5.273}$$

with

$$\theta_n = x - \tilde{a}_n - a_n\tau_1 - \delta_n, \qquad n = 1, 2, \ldots, N, \tag{5.274}$$

$$a_n = \lambda(1 - \gamma_n \cot \gamma_n), \qquad 0 < \gamma_n < \pi, \quad n = 1, 2, \ldots, N, \tag{5.275}$$

$$\tilde{a}_n = \tfrac{3}{4}a_n^2 + \tfrac{3}{4}\lambda a_n - \tfrac{1}{4}(\lambda\gamma_n)^2, \tag{5.276}$$

$$e^{A_{lm}} = \frac{(a_l - a_m)^2 + \lambda^2(\gamma_l - \gamma_m)^2}{(a_l - a_m)^2 + \lambda^2(\gamma_l + \gamma_m)^2}, \tag{5.277}$$

where γ_n and δ_n are real constants. It may be shown that (5.273) reduces to the N-soliton solution of the first higher-order BO equation in the deep-water limit ($\lambda \to \infty$) and to that of the first higher-order KdV equation in the shallow-water limit ($\lambda \to 0$) [29].

5.4 Higher-Order Modified Korteweg–de Vries Equations

5.4.1 Bilinearization

The Lax hierarchy of the modified KdV equation is given by

$$u_t = -(\partial/\partial x)(\delta I_n/\delta u) = -\partial K_n/\partial x, \qquad n = 3, 4, \ldots, \tag{5.278}$$

where I_n is the nth conserved quantity of the modified KdV equation.

The first few K_n derived from the functional derivative of I_n are

$$K_3 = u_{2x} + 2u^3, \tag{5.279}$$

$$K_4 = u_{4x} + 10u(u_x)^2 + 10u^2u_{2x} + 6u^5, \tag{5.280}$$

$$\begin{aligned} K_5 = {} & u_{6x} + 14u^2u_{4x} + 56uu_xu_{3x} + 42u(u_{2x})^2 + 70(u_x)^2u_{2x} \\ & + 70u^4u_{2x} + 140u^3(u_x)^2 + 20u^7, \end{aligned} \tag{5.281}$$

$$\begin{aligned} K_6 = {} & u_{8x} + 18u^2u_{6x} + 108uu_xu_{5x} + 228uu_{2x}u_{4x} + 210(u_x)^2u_{4x} \\ & + 126u^4u_{2x} + 138u(u_{3x})^2 + 756u_xu_{2x}u_{3x} + 1008u^3u_xu_{3x} \\ & + 182(u_{2x})^3 + 756u^3(u_{2x})^2 + 3108u^2(u_x)^2u_{2x} + 420u^6u_{2x} \\ & + 798u(u_x)^4 + 1260u^5(u_x)^2 + 70u^9. \end{aligned} \tag{5.282}$$

For $n = 3$, (5.278) and (5.279) yield the modified KdV equation

$$u_t = -u_{3x} - 6u^2u_x. \tag{5.283}$$

For $n \geq 4$, (5.278) gives the higher-order versions of the modified KdV equation. To bilinearize (5.283), we introduce the dependent variable transformation

$$u = i\frac{\partial}{\partial x}\ln\frac{f'}{f}. \tag{5.284}$$

It may then be seen, by noting formulas (App. I.3.1)–(App. I.3.3), that the coupled bilinear equations [55]

$$D_t f' \cdot f = -D_x^3 f' \cdot f, \tag{5.285}$$

$$D_x^2 f' \cdot f - 0 \tag{5.286}$$

are equivalent to (5.283).

For the purpose of bilinearizing the nth-order modified KdV equation, we first impose $n - 3$ subsidiary conditions for u:

$$u_{\tau_j} = -\partial K_{j+2}/\partial x, \qquad j = 1, 2, \ldots, n-3, \tag{5.287}$$

where τ_j $(j = 1, 2, \ldots, n-3)$ are independent auxiliary variables. Equations (5.278) and (5.287) reduce, by the dependent variable transformation (5.284), to

$$(iD_t f' \cdot f)/f'f = -K_n, \qquad n = 3, 4, \ldots, \tag{5.288}$$

$$(iD_{\tau_j} f' \cdot f)/f'f = -K_{j+2}, \qquad j = 1, 2, \ldots, n-3, \tag{5.289}$$

respectively. Under conditions (5.287), the nth-order modified KdV equation may be transformed into coupled bilinear equations for f' and f [30]:

$$D_t f' \cdot f = \alpha^{(n)} D_x^{2n-3} f' \cdot f + \sum_{j=1}^{n-3} \beta_j^{(n)} D_{\tau_j} D_x^{2(n-j-2)} f' \cdot f, \qquad n \geq 4, \tag{5.290}$$

$$D_x^2 f' \cdot f = 0, \tag{5.291}$$

where $\alpha^{(n)}$ and $\beta_j^{(n)}$ ($j = 1, 2, \ldots, n-3$) are unknown coefficients.

Now comparing the $\varepsilon^{2(n-j-2)}$ terms on both sides of identity (5.19) with $t = \tau_j$, we obtain

$$\begin{aligned} D_{\tau_j} D_x^{2(n-j-2)} f' \cdot f = & \sum_{m=0}^{n-j-2} {}_{2(n-j-2)}C_{2m} (D_x^{2m} f' \cdot f) \frac{\partial}{\partial \tau_j} \\ & \times \frac{\partial^{2(n-j-m-2)}}{\partial x^{2(n-j-m-2)}} \ln \frac{f'}{f} \\ & + \sum_{m=0}^{n-j-3} {}_{2(n-j-2)}C_{2m+1} (D_x^{2m+1} f' \cdot f) \frac{\partial}{\partial \tau_j} \\ & \times \frac{\partial^{2(n-j-m-3)+1}}{\partial x^{2(n-j-m-3)+1}} \ln f'f. \end{aligned} \tag{5.292}$$

Equations (5.291), (5.284), and the definition of the D operator yield

$$\frac{D_x^2 f' \cdot f}{f'f} \equiv \frac{\partial^2}{\partial x^2} \ln f'f + \left[\frac{\partial}{\partial x} \ln \frac{f'}{f} \right]^2 = \frac{\partial^2}{\partial x^2} \ln f'f - u^2 = 0, \tag{5.293}$$

from which we obtain the relation

$$\frac{\partial^2}{\partial x^2} \ln f'f = u^2. \tag{5.294}$$

It follows from (5.287) and (5.294) that

$$\frac{\partial}{\partial \tau_j} \frac{\partial}{\partial x} \ln f'f = 2 \int_{-\infty}^{x} u u_{\tau_j} \, dx' = -2 \int_{-\infty}^{x} u \frac{\partial K_{j+2}}{\partial x'} \, dx' \tag{5.295}$$

and from the translation invariance of I_n that

$$0 = \frac{\partial}{\partial \varepsilon} I_n[u(x + \varepsilon)] \bigg|_{\varepsilon=0} = \int_{-\infty}^{\infty} \frac{\delta I_n}{\delta u} u_{x'} \, dx' = - \int_{-\infty}^{\infty} (K_n)_{x'} u \, dx', \tag{5.296}$$

which implies that $(K_n)_x u$ is a perfect derivative, that is,

$$(K_n)_x u \equiv (J_n)_x. \tag{5.297}$$

Substituting (5.289), (5.295), and (5.297) into (5.292) and dividing both sides by $f'f$, we obtain

$$\begin{aligned} \frac{D_{\tau_j} D_x^{2(n-j-2)} f' \cdot f}{f'f} = i & \sum_{m=0}^{n-j-2} {}_{2(n-j-2)}C_{2m} A_{2m} (K_{j+2})_{2(n-j-m-2)x} \\ & - 2 \sum_{m=0}^{n-j-3} {}_{2(n-j-2)}C_{2m+1} A_{2m+1} (J_{j+2})_{2(n-j-m-3)x}, \end{aligned} \tag{5.298}$$

where we have set

$$A_n = (D_x^n f' \cdot f)/f'f, \qquad n = 0, 1, \ldots. \tag{5.299}$$

The expressions for J_n defined by (5.297) are derived, using (5.279)–(5.281), as

$$J_3 = uu_{2x} - \tfrac{1}{2}(u_x)^2 + \tfrac{3}{2}u^4, \tag{5.300}$$

$$J_4 = uu_{4x} + u_x u_{3x} + 10u^3 u_{2x} - \tfrac{1}{2}(u_{2x})^2 + 5u^2(u_x)^2 + 5u^6, \tag{5.301}$$

$$\begin{aligned} J_5 = {} & uu_{6x} - u_x u_{5x} + u_{2x} u_{4x} + 14u^3 u_{4x} - \tfrac{1}{2}(u_{3x})^2 + 42u^2 u_x u_{3x} \\ & + 49u^2(u_{2x})^2 + 42u(u_x)^2 u_{2x} + 70u^5 u_{2x} - \tfrac{21}{2}(u_x)^4 \\ & + 105u^4(u_x)^2 + \tfrac{35}{2}u^8. \end{aligned} \tag{5.302}$$

To express A_n in terms of $u, u_x, \ldots$, substitute (5.284) and (5.294) into identity (5.19) with $t = x$ and compare the ε^{2n-1} terms and the ε^{2n}

terms on both sides of the resultant equation. Then we find the recursion formulas for A_{2n} and A_{2n+1} as

$$A_{2n} = \frac{1}{i} \sum_{m=0}^{n-1} {}_{2n-1}C_{2m} A_{2(n-m)-1} u_{2mx} + \sum_{m=0}^{n-1} {}_{2n-1}C_{2m+1} A_{2(n-m-1)} (u^2)_{2mx}, \tag{5.303}$$

$$A_{2n+1} = \frac{1}{i} \sum_{m=0}^{n} {}_{2n}C_{2m} A_{2(n-m)} u_{2mx} + \sum_{m=0}^{n-1} {}_{2n}C_{2m+1} A_{2(n-m)-1} (u^2)_{2mx}. \tag{5.304}$$

These formulas are solved successively starting with $A_0 = 1$ and $A_1 = -iu$. Several explicit forms of A_n are given as

$$A_0 = 1, \tag{5.305}$$

$$A_1 = -iu, \tag{5.306}$$

$$A_2 = 0 \qquad \text{by (5.291)}, \tag{5.307}$$

$$A_3 = -i(u_{2x} + 2u^3) = -iK_3 \qquad \text{by (5.279)}, \tag{5.308}$$

$$A_4 = -2uu_{2x} + 2(u_x)^2 - 2u^4, \tag{5.309}$$

$$\begin{aligned} A_5 &= -i[u_{4x} + 10u^2u_{2x} + 10u(u_x)^2 + 6u^5] \\ &= -iK_4 \qquad \text{by (5.280)}, \end{aligned} \tag{5.310}$$

$$A_6 = -4uu_{4x} + 8u_xu_{3x} - 4(u_{2x})^2 - 40u^3u_{2x} - 16u^6, \tag{5.311}$$

$$\begin{aligned} A_7 &= -i[u_{6x} + 14u^2u_{4x} + 56uu_xu_{3x} + 42u(u_{2x})^2 + 70(u_x)^2u_{2x} \\ &\qquad + 70u^4u_{2x} + 140u^3(u_x)^2 + 20u^7] \\ &= -iK_5 \qquad \text{by (5.281)}, \end{aligned} \tag{5.312}$$

$$\begin{aligned} A_8 = {} & -6uu_{6x} + 12u_xu_{5x} - 26u_{2x}u_{4x} - 112u^3u_{4x} - 420u^2(u_{2x})^2 \\ & - 280u(u_x)^2u_{2x} - 616u^5u_{2x} - 280u^4(u_x)^2 + 140(u_x)^4 \\ & + 20(u_{3x})^2 - 132u^8, \end{aligned} \tag{5.313}$$

$$\begin{aligned} A_9 = {} & -i[u_{8x} + 18u^2u_{6x} + 108uu_xu_{5x} + 186uu_{2x}u_{4x} + 252(u_x)^2u_{4x} \\ & + 84u^4u_{4x} + 180u(u_{3x})^2 + 672u_xu_{2x}u_{3x} + 1344u^3u_xu_{3x} \\ & + 224(u_{2x})^3 + 588u^3(u_{2x})^3 + 2520u^2(u_x)^2u_{2x} + 168u^6u_{2x} \\ & + 1260u(u_x)^4 + 1512u^5(u_x)^2 + 28u^9]. \end{aligned} \tag{5.314}$$

Note that

$$A_{2n-3} = -iK_n, \tag{5.315}$$

for $n = 3, 4, 5$, but this relation does not hold generally for $n \geq 6$. Introducing (5.288) and (5.298) into (5.290) we obtain

$$K_n = -i\alpha^{(n)}A_{2n-3} + \sum_{j=1}^{n-3} \beta_j^{(n)}\left[\sum_{m=0}^{n-j-2} {}_{2(n-j-2)}C_{2m}A_{2m}(K_{j+2})_{2(n-j-m-2)x}\right.$$
$$\left. + 2i\sum_{m=0}^{n-j-3} {}_{2(n-j-2)}C_{2m}A_{2m+1}(J_{j+2})_{2(n-j-m-3)x}\right], \qquad n \geq 4. \tag{5.316}$$

Equating the coefficients of $u^{2n-3}, u^{2n-6}u_{2x}, \ldots, u_{2(n-2)x}$ on both sides of (5.316), we can derive simultaneous linear algebraic equations for unknown coefficients $\alpha^{(n)}$ and $\beta_j^{(n)}$ ($j = 1, 2, \ldots, n-3$).

We shall now apply the method of bilinearization, as developed so far, to the higher-order modified KdV equations (5.278) with $n = 4, 5, 6$ and present the explicit N-soliton solutions for these equations.

(i) $n = 4$

It follows from (5.316) with $n = 4$ that

$$K_4 = -i\alpha^{(4)}A_5 + \beta_1^{(4)}\left[\sum_{m=0}^{1} {}_2C_{2m}A_{2m}(K_3)_{2(1-m)x} + 4iA_1J_3\right]. \tag{5.317}$$

Substituting (5.279), (5.280), (5.300), (5.305)–(5.307), and (5.310) into (5.317) yields

$$\begin{aligned} u_{4x} + 10u(u_x)^2 + 10u^2u_{2x} + 6u^5 &= -(\alpha^{(4)} - \beta_1^{(4)})u_{4x} - 10(\alpha^{(4)} - \beta_1^{(4)})u(u_x)^2 \\ &\quad - 10(\alpha^{(4)} - \beta_1^{(4)})u^2u_{2x} - 6(\alpha^{(4)} - \beta_1^{(4)})u^5. \end{aligned} \tag{5.318}$$

Comparing the coefficients of u_{4x}, $u(u_x)^2$, u^2u_{2x}, and u^5 on both sides of (5.318), we obtain only one independent equation for two unknown coefficients $\alpha^{(4)}$ and $\beta_1^{(4)}$,

$$-(\alpha^{(4)} - \beta_1^{(4)}) = 1, \tag{5.319}$$

from which $\alpha^{(4)}$ and $\beta_1^{(4)}$ are determined as

$$\alpha^{(4)} = -1 + c_1^{(4)}, \tag{5.320}$$

$$\beta_1^{(4)} = c_1^{(4)}, \tag{5.321}$$

where $c_1^{(4)}$ is an arbitrary constant. If we do not introduce the auxiliary variable τ_1, $c_1^{(4)}$ can be taken to be zero.

(ii) $n = 5$

With $n = 5$, it follows from (5.316) that

$$K_5 = -i\alpha^{(5)} + \sum_{j=1}^{2} \beta_j^{(5)} \left[\sum_{m=0}^{3-j} {}_{2(3-j)}C_{2m} A_{2m} (K_{j+2})_{2(3-j-m)x} + 2i \sum_{m=0}^{2-j} {}_{2(3-j)}C_{2m+1} A_{2m+1} (J_{j+2})_{2(2-j-m)x} \right]. \tag{5.322}$$

Substituting (5.279)–(5.281), (5.300), (5.301), (5.305)–(5.309), and (5.312) into (5.322) and comparing the coefficients of $u_{6x}, u^2u_{4x}, \ldots, u^7$ on both sides of (5.322) yields eight equations, but essentially one independent equation for three unknown coefficients $\alpha^{(5)}$, $\beta_1^{(5)}$, and $\beta_2^{(5)}$ given by

$$-\alpha^{(5)} + \beta_1^{(5)} + \beta_2^{(5)} = 1, \tag{5.323}$$

which has been obtained by comparing the coefficient of u_{6x}. Thus, $\alpha^{(5)}$, $\beta_1^{(5)}$, and $\beta_2^{(5)}$ are determined as

$$\alpha^{(5)} = -1 + c_1^{(5)} + c_2^{(5)}, \tag{5.324}$$

$$\beta_1^{(5)} = c_1^{(5)}, \tag{5.325}$$

$$\beta_2^{(5)} = c_2^{(5)}, \tag{5.326}$$

where $c_1^{(5)}$ and $c_2^{(5)}$ are arbitrary constants. If we set $c_1^{(5)} = c_2^{(5)} = 0$, we need not introduce the auxiliary variables τ_1 and τ_2.

(iii) $n = 6$

With $n = 6$, it follows from (5.136) that

$$K_6 = -i\alpha^{(6)} A_9 + \sum_{j=1}^{3} \beta_j^{(6)} \left[\sum_{m=0}^{4-j} {}_{2(4-j)}C_{2m} A_{2m} (K_{j+2})_{2(4-j-m)x} + 2i \sum_{m=0}^{3-j} {}_{2(4-j)}C_{2m+1} A_{2m+1} (J_{j+2})_{2(3-j-m)x} \right]. \tag{5.327}$$

Substituting (5.279)–(5.281), (5.300)–(5.303), (5.305)–(5.310), and (5.313) into (5.327) and comparing the coefficients of $u_{8x}, u^2u_{6x}, \ldots, u^9$ on both sides of (5.327) yields formally 16 equations, but essentially two independent equations for four unknown coefficients $\alpha^{(6)}$, $\beta_1^{(6)}$, $\beta_2^{(6)}$, and $\beta_3^{(6)}$ given by

$$-\alpha^{(6)} + \beta_1^{(6)} + \beta_2^{(6)} + \beta_3^{(6)} = 1, \tag{5.328}$$

$$-28\alpha^{(6)} + 76\beta_1^{(6)} + 68\beta_2^{(6)} + 70\beta_3^{(6)} = 70. \tag{5.329}$$

Equations (5.328) and (5.329) have been obtained by comparing the coefficients of u_{8x} and u^9, respectively. Thus, $\alpha^{(6)}$, $\beta_1^{(6)}$, $\beta_2^{(6)}$, and $\beta_3^{(6)}$ are determined as

$$\alpha^{(6)} = -\tfrac{1}{8} + \tfrac{1}{6}c_1^{(6)} + \tfrac{1}{8}c_2^{(6)}, \tag{5.330}$$

$$\beta_1^{(6)} = \tfrac{7}{8} - \tfrac{5}{6}c_1^{(6)} - \tfrac{7}{8}c_2^{(6)}, \tag{5.331}$$

$$\beta_2^{(6)} = c_1^{(6)}, \tag{5.332}$$

$$\beta_3^{(6)} = c_2^{(6)}, \tag{5.333}$$

where $c_1^{(6)}$ and $c_2^{(6)}$ are arbitrary constants. If we choose $c_1^{(6)} = c_2^{(6)} = 0$, we need not introduce the auxiliary variables τ_2 and τ_3. Note that, if we set $c_1^{(4)} = c_2^{(5)} = c_3^{(5)} = 0$ in (5.320), (5.321), and (5.324)–(5.326), then for $n = 4, 5$ the higher-order modified KdV equations can be transformed, without introducing the auxiliary variables τ_1 and τ_2, into the coupled bilinear equations

$$D_t f' \cdot f = -D_x^{2n-3} f' \cdot f, \tag{5.334}$$

$$D_x^2 f' \cdot f = 0. \tag{5.335}$$

These bilinear equations include the bilinear equations (5.285) and (5.286) of the original modified KdV equation as a special case of $n = 3$. However, for $n \geq 6$, we cannot generally transform the higher-order modified KdV equations into such simple forms as (5.334) and (5.335). In the case of the higher-order KdV equations, even the first higher-order equation cannot be transformed into the bilinear equation without introducing an auxiliary variable, as shown in Section 5.2. This is an interesting difference between the higher-order KdV equations and the higher-order modified KdV equations. Since the modified KdV equation is related to the KdV equation by the Bäcklund transformation [105], it seems worthwhile to study the relation between these higher-order equations from the viewpoint of the Bäcklund transformation, which will be shown in the next section.

5.4.2 *Solutions*

The N-soliton solutions of the nth-order modified KdV equation with $n = 3, 4, 5, 6$ are expressed compactly in the forms

$$f = \sum_{\mu=0,1} \left[\sum_{l=1}^{N} \mu_l \left(\theta_l + \frac{\pi i}{2} \right) + \sum_{l<m}^{(N)} \mu_l \mu_m A_{lm} \right], \tag{5.336}$$

$$f' = \sum_{\mu=0,1} \left[\sum_{l=1}^{N} \mu_l \left(\theta_l - \frac{\pi i}{2} \right) + \sum_{l<m}^{(N)} \mu_l \mu_m A_{lm} \right], \tag{5.337}$$

with

$$\theta_l = a_l \left[x - a_l^{2(n-2)} t - \sum_{j=1}^{n-3} a_l^{2j} \tau_j - \delta_l \right], \qquad l = 1, 2, \ldots, N, \tag{5.338}$$

$$e^{A_{lm}} = (a_l - a_m)^2/(a_l + a_m)^2, \qquad l, m = 1, 2, \ldots, N, \tag{5.339}$$

where a_l and δ_l are arbitrary constants.

If we set

$$\tilde{\delta}_l = \delta_l + \sum_{j=1}^{n-3} a_l^{2j} \tau_j, \tag{5.340}$$

the form of the solution is the same for all equations, the only difference being the velocity $a_l^{2(n-2)}$ of solitons. This is a common property of the solutions of the Lax hierarchy of nonlinear evolution equations.

5.5 Bäcklund Transformations and Inverse Scattering Transforms of Higher-Order Korteweg–de Vries Equations

5.5.1 *Bäcklund Transformations*

In this section we shall derive the Bäcklund transformation of the Lax hierarchy of the KdV equation, which is written as

$$u_t = -\partial K_n(u)/\partial x, \qquad n = 3, 4, \ldots, \tag{5.341}$$

on the basis of the bilinear transformation method [31]. As shown in Section 5.2, Eqs. (5.341) are transformed into bilinear equations through the dependent variable transformation

$$u = 2\frac{\partial^2}{\partial x^2}\ln f. \tag{5.342}$$

Denote another solution of (5.341) as

$$u' = 2\frac{\partial^2}{\partial x^2}\ln f' \tag{5.343}$$

and assume that f' also satisfies the spatial part of the Bäcklund transformation

$$D_x^2 f'\cdot f = \lambda f'f, \tag{5.344}$$

where λ is an arbitrary parameter. Furthermore, introduce the quantities

$$w_x = u, \tag{5.345}$$

$$w'_x = u', \tag{5.346}$$

$$v = i\frac{\partial}{\partial x}\ln\frac{f'}{f}. \tag{5.347}$$

Using (5.342), (5.343), and (5.345)–(5.347), (5.344) is rewritten as

$$\frac{D_x^2 f'\cdot f}{f'f} \equiv \frac{\partial^2}{\partial x^2}\ln f'f + \left(\frac{\partial}{\partial x}\ln\frac{f'}{f}\right)^2 = \frac{1}{2}(w'_x + w_x) - v^2 = \lambda, \tag{5.348}$$

which leads to the relation

$$w'_x + w_x = 2v^2 + 2\lambda. \tag{5.349}$$

On the other hand, it follows from (5.342), (5.343), and (5.345)–(5.347) that

$$w'_x - w_x = u' - u = 2\frac{\partial^2}{\partial x^2}\ln\frac{f'}{f} = -2iv_x, \tag{5.350}$$

which, combined with (5.349), yields

$$w'_x = -iv_x + v^2 + \lambda, \tag{5.351}$$

$$w_x = iv_x + v^2 + \lambda. \tag{5.352}$$

To derive the time part of the Bäcklund transformation, substitute (5.345) and (5.346) into (5.341), integrate with respect to x, and use the boundary condition, $K(u) \to 0$ as $|x| \to \infty$. Then we obtain the relation

$$w'_t - w_t = -[K_n(w'_x) - K_n(w_x)]. \tag{5.353}$$

On the other hand, it follows from (5.342), (5.343), (5.345), and (5.346) that

$$w'_t - w_t = 2\frac{\partial^2}{\partial t\, \partial x} \ln \frac{f'}{f} = 2\frac{\partial}{\partial x}\frac{D_t f' \cdot f}{f' f}. \tag{5.354}$$

Equating (5.353) and (5.354) and substituting (5.351) and (5.352) into the resultant equation, we obtain

$$2\frac{\partial}{\partial x}\frac{D_t f' \cdot f}{f' f} = -[K_n(-iv_x + v^2 + \lambda) - K_n(iv_x + v^2 + \lambda)]. \tag{5.355}$$

We now introduce the relation

$$K_n(\pm iv_x + v^2 + \lambda) = \sum_{m=0}^{n-1} C_{n,m} K_{n-m}(\pm iv_x + v^2)\lambda^m, \tag{5.356}$$

where

$$C_{n,m} = \frac{(2n-2)!(n-m-1)!}{m!(n-1)!(2n-2m-2)!}. \tag{5.357}$$

This relation is derived by the formula [54]

$$\int_{-\infty}^{\infty} \frac{\delta I_n}{\delta u}\, dx = 2(2n-3)I_{n-1}, \qquad n = 2, 3, \ldots, \tag{5.358}$$

as follows: Repeated use of (5.358) and the definition of K_n $(= \delta I_n / \delta u)$ gives

$$\int_{-\infty}^{\infty} \cdots \int_{-\infty}^{\infty} \frac{\delta^m K_n(u)}{\delta u(x_1, t) \cdots \delta u(x_m, t)}\, dx_1 \cdots dx_m$$
$$= \frac{(2n-2)!(n-m-1)!}{(n-1)!(2n-2m-2)!} K_{n-m}(u). \tag{5.359}$$

On the other hand, since $K_n(u)$ is expressed in the form

$$K_n(u) = u_{2(n-2)x} + \cdots + cu^{n-1}, \qquad c \neq 0, \tag{5.360}$$

we obtain from (5.359)

$$K_n(u+\lambda) = K_n(u) + \sum_{m=1}^{n-1} \frac{\lambda^m}{m!} \times \int_{-\infty}^{\infty} \cdots \int_{-\infty}^{\infty} \frac{\delta^m K_n(u)}{\delta u(x_1, t) \cdots \delta u(x_m, t)} dx_1 \cdots dx_m$$

$$= K_n(u) + \sum_{m=1}^{n-1} \lambda^m \frac{(2n-2)!(n-m-1)!}{m!(n-1)!(2n-2m-2)!} K_{n-m}(u)$$

$$= \sum_{m=0}^{n-1} C_{n,m} \lambda^m K_{n-m}(u). \tag{5.361}$$

Moreover, we note the result from Adler and Moser [116]

$$\frac{\partial}{\partial x} K_n(\pm i v_x + v^2) = \left(\pm i \frac{\partial}{\partial x} + 2v\right) \frac{\partial}{\partial x} \tilde{K}_n(v), \tag{5.362}$$

where $\tilde{K}_n$ is derived from the functional derivative of the nth conserved quantity $\tilde{I}_n$ of the modified KdV equation

$$\tilde{K}_n(v) = \delta \tilde{I}_n / \delta v, \tag{5.363}$$

explicit forms of which have been given in (5.279)–(5.282). It should be noted that

$$\tilde{K}_n(-v) = -\tilde{K}_n(v). \tag{5.364}$$

Introducing (5.356) into (5.355) and using (5.362), we obtain, after integration with respect to x,

$$D_t f' \cdot f / f' f = i \sum_{m=0}^{n-2} C_{n,m} \lambda^m \tilde{K}_{n-m}(v), \tag{5.365}$$

where the relation K_1 = constant has been used and the integration constant has been taken to be zero. It is important to observe that (5.365) reduces to the nth order modified KdV equation in the limit of $\lambda \to 0$

$$v_t = -\partial \tilde{K}_n(v)/\partial x, \qquad n = 3, 4, \ldots. \tag{5.366}$$

The Bäcklund transformation (5.349) and (5.365) is rewritten in terms of the original variables w and w' as [117]

$$w'_x + w_x = -\tfrac{1}{2}(w' - w)^2 + 2\lambda, \tag{5.367}$$

$$w'_t - w_t = 2i \sum_{m=0}^{n-2} C_{n,m} \lambda^m (\partial/\partial x) \tilde{K}_{n-m}[\tfrac{1}{2}i(w' - w)], \tag{5.368}$$

which follow from (5.349), (5.365), (5.342), (5.343), and (5.345)–(5.347). Equations (5.367) and (5.368) with $n = 3$ correspond to the Bäcklund transformation of the KdV equation (2.163) and (2.166), first developed by Wahlquist and Estabrook [50]. To express the right-hand side of (5.365) in bilinear form, we must bilinearize $K_{n-m}(v)$ under condition (5.344), following the procedure developed for the higher-order modified KdV equations (see Section 5.4). As shown in Section 5.4, the nth-order modified KdV equation (5.366) is transformed into the coupled bilinear equations

$$D_t f' \cdot f = \alpha^{(n)} D_x^{2n-3} f' \cdot f + \sum_{j=1}^{n-3} \beta_j^{(n)} D_{\tau_j} D_x^{2(n-j-2)} f' \cdot f, \tag{5.369}$$

$$D_x^2 f' \cdot f = 0, \tag{5.370}$$

under the $n - 3$ subsidiary conditions

$$D_{\tau_s} f' \cdot f = \alpha^{(s+2)} D_x^{2s+1} f' \cdot f + \sum_{j=1}^{s-1} \beta_j^{(s+2)} D_{\tau_j} D_x^{2(s-j)} f' \cdot f, \qquad s = 1, 2, \ldots, n-3, \tag{5.371}$$

through the dependent variable transformation (5.347). Substitution of (5.366) and (5.347) into (5.369) gives

$$\tilde{K}_n(v) = -i\alpha^{(n)} (D_x^{2n-3} f' \cdot f)/f'f - i \sum_{j=1}^{n-3} \beta_j^{(n)} [D_{\tau_j} D_x^{2(n-j-2)} f' \cdot f]/f'f. \tag{5.372}$$

We now define the quantities

$$A_n(\lambda) = (D_x^n f' \cdot f)/f'f, \qquad A_2 = \lambda \quad \text{by (5.344)}, \tag{5.373}$$

$$A_{j,n}(\lambda) = (D_{\tau_j} D_x^n f' \cdot f)/f'f, \tag{5.374}$$

$$\tilde{A}_{j,n}(\lambda) = \frac{\partial}{\partial \tau_j} \frac{\partial^n}{\partial x^n} \ln \frac{f'}{f}, \tag{5.375}$$

$$B_{j,n}(\lambda) = \frac{\partial}{\partial \tau_j} \frac{\partial^n}{\partial x^n} \ln f'f. \tag{5.376}$$

Using these definitions, (5.372) may be rewritten as

$$\tilde{K}_n(v) = -i\alpha^{(n)}A_{2n-3}(0) - i\sum_{j=1}^{n-3}\beta_j^{(n)}A_{j,2(n-j-2)}(0). \tag{5.377}$$

The recursion formulas corresponding to (5.303) and (5.304) are obtained by noting the relation

$$\frac{\partial^2}{\partial x^2}\ln f'f \equiv A_2(\lambda) - \left(\frac{\partial}{\partial x}\ln\frac{f'}{f}\right)^2 = \lambda + v^2, \tag{5.378}$$

which is a consequence of (5.344) and (5.347) by

$$A_{2n}(\lambda) = \frac{1}{i}\sum_{m=0}^{n-1}{}_{2n-1}C_{2m}A_{2(n-m)-1}(\lambda)v_{2mx}$$
$$+\sum_{m=0}^{n-1}{}_{2n-1}C_{2m+1}A_{2(n-m-1)}(\lambda)(v^2+\lambda)_{2mx}, \tag{5.379}$$

$$A_{2n+1}(\lambda) = \frac{1}{i}\sum_{m=0}^{n}{}_{2n}C_{2m}A_{2(n-m)}(\lambda)v_{2mx}$$
$$+\sum_{m-0}^{n-1}{}_{2n}C_{2m+1}A_{2(n-m)-1}(\lambda)(v^2+\lambda)_{2mx}. \tag{5.380}$$

In the limit of $\lambda \to 0$, (5.379) and (5.380) reduce to (5.303) and (5.304), respectively.

To express $A_n(\lambda)$ in terms of $A_n(0)$, we expand $A_n(\lambda)$ in powers of λ as

$$A_n(\lambda) = \sum_{j=0}^{[n/2]}\tilde{a}_j^{(n)}\lambda^j A_{n-2j}(0), \qquad n = 0, 1, 2, \ldots, \tag{5.381}$$

where $\tilde{a}_j^{(n)}$ are coefficients that will be determined later, and we may set

$$\tilde{a}_{n-1}^{(2n)} = 0, \qquad n = 1, 2, \ldots, \tag{5.382}$$

since $A_2(0) = 0$ by (5.344). In (5.381) the notation $[n/2]$ means the maximum integer that does not exceed $n/2$. Then it is straightforward to derive the recursion formulas for $a_j^{(n)}$ as

$$\tilde{a}_j^{(n)} = (n/n-2j)\tilde{a}_j^{(n-1)} \qquad \text{for} \quad 0 \leq j \leq [n/2]-1, \tag{5.383}$$

$$\tilde{a}_n^{(2n)} = (2n-1)\tilde{a}_{n-1}^{(2n-2)}. \tag{5.384}$$

These relations follow by substituting (5.381) into (5.379) and (5.380) and comparing the coefficients of λ^j on both sides of the resultant equations. The solution of (5.383) and (5.384) may be obtained as

$$\tilde{a}_j^{(n)} = n!/2^j(n-2j)!j!, \qquad 0 \le j \le [n/2], \tag{5.385}$$

with an initial condition $\tilde{a}_1^{(2)} = 1$, which is derived from the relation $A_2(\lambda) = \lambda$. Substituting (5.385) into (5.381) and solving these equations with respect to $A_n(0)$, we obtain

$$A_n(0) = \sum_{j=0}^{[n/2]} a_j^{(n)} \lambda^j A_{n-2j}(\lambda), \tag{5.386}$$

with

$$a_j^{(n)} = \frac{(-1)^j n!}{2^j(n-2j)!j!} = (-1)^j \tilde{a}_j^{(n)}. \tag{5.387}$$

Now Eq. (5.292) is rewritten in terms of (5.373)–(5.376) as

$$\begin{aligned} A_{j,2n}(0) = {} & \sum_{m=0}^{n} {}_{2n}C_{2m} A_{2m}(0)\tilde{A}_{j,2(n-m)}(0) \\ & + \sum_{m=0}^{n-1} {}_{2n}C_{2m+1} A_{2m+1}(0) B_{j,2(n-m-1)}(0). \end{aligned} \tag{5.388}$$

However

$$B_{j,n}(\lambda) = \frac{\partial}{\partial \tau_j}\frac{\partial^n}{\partial x^n} \ln f'f = \frac{\partial}{\partial \tau_j}\frac{\partial^{n-2}}{\partial x^{n-2}}(v^2 + \lambda) = \frac{\partial}{\partial \tau_j}\frac{\partial^{n-2}}{\partial x^{n-2}} v^2 = B_{j,n}(0), \tag{5.389}$$

and

$$\tilde{A}_{j,n}(\lambda) = \tilde{A}_{j,n}(0), \tag{5.390}$$

by (5.375), the definition of $\tilde{A}_{j,n}$. Substituting (5.386), (5.389), and (5.390) into (5.388) yields, after some algebra,

$$\begin{aligned} A_{j,2n}(0) &= \sum_{s=1}^{n} \lambda^{n-s} \frac{(-1)^{n-s}(2n)!}{2^{n-s}(n-s)!(2s)!} \left[\sum_{m=0}^{s} {}_{2s}C_{2m} A_{2m}(\lambda) \tilde{A}_{j,2(s-m)}(\lambda) \right. \\ &\quad \left. + \sum_{m=0}^{s-1} {}_{2s}C_{2m+1} A_{2m+1}(\lambda) B_{j,2(s-m-1)}(\lambda) \right] + \lambda^n a_n^{(2n)} \tilde{A}_{j,0}(\lambda) \\ &= \sum_{s=0}^{n} \lambda^s a_s^{(2n)} A_{j,2(n-s)}(\lambda). \end{aligned} \tag{5.391}$$

Introducing (5.386) and (5.391) into (5.377), we obtain

$$\tilde{K}_n(v) = -i\alpha^{(n)} \sum_{j=0}^{n-2} a_j^{(2n-3)} \lambda^j A_{2n-2j-3}(\lambda) - i \sum_{j=1}^{n-3} \beta_j^{(n)} \sum_{s=0}^{n-j-2} a_s^{(2n-2j-4)} \lambda^s A_{j,2(n-j-s-2)}(\lambda). \quad (5.392)$$

Finally, substituting (5.392) into (5.365) and using the definition of A_n and $A_{j,n}$, we obtain

$$\begin{aligned} D_t f' \cdot f = {} & \sum_{m=0}^{n-4} \lambda^m C_{n,m} \Bigg(\alpha^{(n-m)} \sum_{j=0}^{n-m-2} a_j^{(2n-2m-3)} \lambda^j D_x^{2(n-m-j)-3} f' \cdot f \\ & + \sum_{j=1}^{n-m-3} \beta_j^{(n-m)} \sum_{s=0}^{n-m-j-2} a_s^{(2n-2m-2j-4)} \lambda^s \\ & \times D_{\tau_j} D_x^{2(n-m-j-s-2)} f' \cdot f \Bigg) \\ & - C_{n,n-3} \lambda^{n-3} D_x^3 f' \cdot f - (C_{n,n-2} - 3C_{n,n-3}) \lambda^{n-2} D_x f' \cdot f. \end{aligned} \quad (5.393)$$

The subsidiary conditions corresponding to (5.371) are expressed as

$$D_{\tau_l} f' \cdot f = \text{right-hand side of (5.393) with } n \text{ replaced by } l+2, \quad l = 1, 2, \ldots, n-3. \quad (5.394)$$

Equations (5.344), (5.393), and (5.394) constitute the Bäcklund transformation of the nth-order KdV equation expressed in bilinear form.

For $n = 4$, Eqs. (5.393) and (5.394) give

$$\begin{aligned} D_t f' \cdot f = {} & \alpha^{(4)} D_x^5 f' \cdot f - 10(\alpha^{(4)} + 1)\lambda D_x^3 f' \cdot f + 15\alpha^{(4)} \lambda^2 D_x f' \cdot f \\ & + \beta_1^{(4)} (D_{\tau_1} D_x^2 f' \cdot f - \lambda D_{\tau_1} f' \cdot f), \end{aligned} \quad (5.395)$$

$$D_{\tau_1} f' \cdot f = -D_x^3 f' \cdot f - 3\lambda D_x f' \cdot f, \quad (5.396)$$

where $\alpha^{(4)}$ and $\beta_1^{(4)}$ are given by (5.320) and (5.321), respectively.

Finally, we shall derive the conserved quantity of the nth-order KdV equation. For this purpose, expand $w' - w$ as

$$w' - w = 2\eta + \sum_{m=1}^{\infty} f_m^{(n)} \eta^{-m}, \qquad \eta \equiv \lambda^{1/2}, \quad (5.397)$$

and substitute (5.397) into (5.367). The result is expressed in the form of the recursion formula as

$$f_1^{(n)} = -u, \tag{5.398}$$

$$f_2^{(n)} = \frac{1}{2} u_x, \tag{5.399}$$

$$f_{m+1}^{(n)} = -\frac{1}{2}(f_m^{(n)})_x - \frac{1}{4}\sum_{s=1}^{m-1} f_s^{(n)} f_{m-s}^{(n)}, \qquad m \geq 2. \tag{5.400}$$

Substituting (5.397) into (5.368), integrating with respect to x, and using the boundary condition, we obtain

$$\frac{\partial}{\partial t}\int_{-\infty}^{\infty} f_m^{(n)}\,dx = 0, \qquad m = 1, 2, \ldots, \tag{5.401}$$

which implies that

$$I_m^{(n)} = \int_{-\infty}^{\infty} f_m^{(n)}\,dx, \qquad m = 1, 2, \ldots, \tag{5.402}$$

are the conserved quantities of the nth-order KdV equation. We see that the functional form of the conserved quantities $I_m^{(n)}$ is the same for all n, since the spatial part of the Bäcklund transformation (5.367) does not depend on n. This is a remarkable property of the Lax hierarchy of the KdV equation.

5.5.2 *Inverse Scattering Transforms*

As discussed in Section 2.5, the inverse scattering transform can be derived from the Bäcklund transformation. To show this explicitly in the case of the higher-order KdV equations, define the wave function ψ by

$$\psi = f'/f, \tag{5.403}$$

and note the identity of the bilinear operator

$$\begin{aligned}\exp(\varepsilon D_x + \delta D_{\tau_j}) f' \cdot f \\ = \left\{\exp\left[2\cosh\left(\varepsilon\frac{\partial}{\partial x} + \delta\frac{\partial}{\partial \tau_j}\right)\ln f\right]\right\}\left[\exp\left(\varepsilon\frac{\partial}{\partial x} + \delta\frac{\partial}{\partial \tau_j}\right)\frac{f'}{f}\right],\end{aligned} \tag{5.404}$$

which is derived from (App. I.3). Substituting (5.403) into (5.404) yields

$$\frac{\exp(\varepsilon D_x + \delta D_{\tau_j}) f' \cdot f}{f^2} = \left\{\exp\left[2 \sum_{m=1}^{\infty} \frac{1}{(2m)!} \left(\varepsilon \frac{\partial}{\partial x} + \delta \frac{\partial}{\partial \tau_j}\right)^{2m} \ln f\right]\right\} \exp\left(\varepsilon \frac{\partial}{\partial x} + \delta \frac{\partial}{\partial \tau_j}\right)\psi. \tag{5.405}$$

On the other hand, from the subsidiary condition for u [sec (5.82)],

$$u_{\tau_j} = -\partial K_{j+2}/\partial x, \qquad j = 1, 2, \ldots, n-3, \tag{5.406}$$

and (5.342), we have, after integration with respect to x,

$$2 \frac{\partial^2}{\partial \tau_j \, \partial x} \ln f = -K_{j+2}. \tag{5.407}$$

Introducing (5.342) and (5.407) into (5.405) and comparing the order δ terms on both sides of (5.405), we obtain

$$\begin{aligned} \frac{D_{\tau_j} \exp(\varepsilon D_x) f' \cdot f}{f^2} &= \exp\left[\sum_{m=1}^{\infty} \frac{\varepsilon^{2m}}{(2m)!} u_{2(m-1)x}\right]\left\{\exp\left(\varepsilon \frac{\partial}{\partial x}\right)\frac{\partial \psi}{\partial \tau_j} - \left[\exp\left(\varepsilon \frac{\partial}{\partial x}\right)\psi\right]\right. \\ &\quad \left. \times \sum_{m=1}^{\infty} \frac{\varepsilon^{2m-1}}{(2m-1)!} (K_{j+2})_{2(m-1)x}\right\}. \end{aligned} \tag{5.408}$$

It follows from (5.405) with $\delta = 0$ and (5.342) that

$$\frac{\exp(\varepsilon D_x) f' \cdot f}{f^2} = \exp\left[\sum_{m=1}^{\infty} \frac{\varepsilon^{2m}}{(2m)!} u_{2(m-1)x}\right] \exp\left(\varepsilon \frac{\partial}{\partial x}\right)\psi. \tag{5.409}$$

If we use (5.408) and (5.409) expanded in powers of ε, we can express the right-hand side of (5.393) in terms of $\psi, \psi_x, \ldots, \psi_{\tau_j}, \ldots, u, u_x \ldots$. It is important to note that the resultant expressions are linear with respect to ψ. Eliminating ψ_{τ_j} in (5.393) by using the subsidiary conditions (5.394), we can express (5.393), which describes the time evolution of the inverse scattering transform, in terms of $\psi, \psi_x, \ldots, u, u_x, \ldots$.

The explicit expressions for $n = 4, 5$ are given as

$$\psi_t = -\psi_{5x} - 10u\psi_{3x} - 5(u_{2x} + 3u^2 + 3\lambda^2)\psi_x, \qquad n = 4, \quad (5.410)$$

$$\begin{aligned}\psi_t = &-\psi_{7x} - 7(3u - \lambda)\psi_{5x} - 70(2u^2 - 2\lambda u + \lambda^2)\psi_{3x} \\ &- 7[u_{4x} + 10uu_{2x} + 10(2u^3 - 3\lambda u^2 + 3\lambda^2 u)]\psi_x \\ &- 35[-u_{2x} + (u - \lambda)^2]u_x\psi, \qquad n = 5. \quad (5.411)\end{aligned}$$

The spatial evolution of the inverse scattering transform may be obtained by noting the relation

$$w' - w = 2\psi_x/\psi, \quad (5.412)$$

which derives from (5.342), (5.343), (5.345), (5.346), and (5.403), and substituting (5.412) into (5.367) to yield

$$\psi_{2x} + u\psi = \lambda\psi. \quad (5.413)$$

Thus, we have completed the derivation of the inverse scattering transform from the Bäcklund transformation written in bilinear form. We note that the time evolution of the inverse scattering transform is expressed in a more compact form as

$$\psi_t = i\sum_{m=0}^{n-2} C_{n,m}\lambda^m \tilde{K}_{n-m}(i\psi_x/\psi) + c\psi, \quad (5.414)$$

where c is an integration constant. Equations (5.347), (5.368), and (5.403) have been used in deriving (5.414). It may be seen that Eqs. (5.413) and (5.414) with $n = 3$ correspond to the inverse scattering transform of the KdV equation first derived by Gardner *et al.* [1].

6

Topics Related to the Benjamin–Ono Equation

This chapter is concerned with recent topics related to the BO equation. The equations considered here are the modified BO equation, the derivative nonlinear Schrödinger equation, and the perturbed BO equation. In Section 6.1 the modified BO equation [32] is constructed from the Bäcklund transformation of the BO equation, and the N-soliton and N-periodic wave solutions are presented. The derivative nonlinear Schrödinger equation, briefly discussed in Section 6.2, derives from the nonlinear self-modulation problem of the BO equation; it describes the long-time behavior of the complex amplitude of the basic wave train [34]. Finally, in Section 6.3 the effect of small dissipation on the BO equation is considered. Owing to dissipation, the amplitude of BO solitons decays slowly in time, and a system of equations describing the long-time behaviors of amplitudes and phases of BO solitons are derived by employing the multiple time-scale expansion.

6.1 The Modified Benjamin–Ono Equation

The modified BO equation is generated from the Bäcklund transformation of the BO equation and may be written as [32]

$$w_t + Hw_{xx} + 2\gamma(1 - e^{-w})w_x - w_x Hw_x = 0. \tag{6.1}$$

Equation (6.1) has been obtained from (3.180) by setting

$$2/\varepsilon = \gamma, \tag{6.2}$$

$$\beta = 1. \tag{6.3}$$

It is clear from the construction of the Bäcklund transformation of the BO equation that (6.1) is transformed into the system of bilinear equations

$$(iD_t + 2\gamma D_x - D_x^2 - \mu) f \cdot g = 0, \tag{6.4}$$

$$(iD_t + 2\gamma D_x - D_x^2 - \mu) f' \cdot g' = 0, \tag{6.5}$$

$$(D_x - \gamma i) f \cdot g' = -\gamma i f' g \tag{6.6}$$

by introducing the dependent variable transformation

$$w = \ln(g' f / g f'), \tag{6.7}$$

where $f, f' g$, and g' have the forms

$$f \propto \prod_{j=1}^{N} [x - x_j(t)], \tag{6.8}$$

$$f' \propto \prod_{j=1}^{N'} [x - x_j'(t)], \tag{6.9}$$

$$g \propto \prod_{j=1}^{M} [x - y_j(t)], \tag{6.10}$$

$$g' \propto \prod_{j=1}^{M'} [x - y_j'(t)], \tag{6.11}$$

with the conditions

$$\operatorname{Im} x_j(t) > 0, \qquad j = 1, 2, \ldots, N, \tag{6.12}$$

$$\operatorname{Im} x_j'(t) < 0, \qquad j = 1, 2, \ldots, N', \tag{6.13}$$

$$\operatorname{Im} y_j(t) > 0, \qquad j = 1, 2, \ldots, M, \tag{6.14}$$

$$\operatorname{Im} y_j'(t) < 0, \qquad j = 1, 2, \ldots, M', \tag{6.15}$$

where $x_j(t)$, $x_j'(t)$, $y_j(t)$, and $y_j'(t)$ are complex functions of t, and N, N', M, and M' are positive finite or infinite integers. Equations (6.4)–(6.6) correspond to Eqs. (3.156)–(3.158), respectively. Transformation (6.7) follows from (3.171), (3.165), and (3.166). It is an easy exercise to reconstruct (6.1) from (6.4)–(6.6) and (6.7).

We shall now construct the N-soliton and N-periodic wave solutions of the coupled bilinear equations (6.4)–(6.6). For $N = 1$, it is confirmed by direct substitution that the following pair of solutions are satisfied by (6.4)–(6.6):

$$f = 1 + \exp(i\xi_1 - \phi_1 + \psi_1), \tag{6.16}$$

$$f' = 1 + \exp(i\xi_1 - \phi_1 - \psi_1), \tag{6.17}$$

$$g = 1 + \exp(i\xi_1 + \phi_1 + \psi_1), \tag{6.18}$$

$$g' = 1 + \exp(i\xi_1 + \phi_1 - \psi_1). \tag{6.19}$$

Here

$$\xi_1 = k_1(x - a_1 t - x_{01}) + \xi_1^{(0)}, \tag{6.20}$$

$$\exp(2\psi_1) = (a_1 + k_1)/(a_1 - k_1), \tag{6.21}$$

$$a_1 = k_1 \coth \phi_1 + 2\gamma, \tag{6.22}$$

with k_1, ϕ_1, x_{01}, and $\xi_1^{(0)}$ being real constants. To satisfy (6.12)–(6.15), the following conditions must be imposed on k_1, ϕ_1, and ψ_1:

$$(\psi_1 - \phi_1)/k_1 > 0, \tag{6.23}$$

$$(\psi_1 + \phi_1)/k_1 < 0. \tag{6.24}$$

The present solution may be expressed in terms of the original variable in the form

$$w = \ln \frac{\cos \xi_1 + \cosh(\psi_1 - \phi_1)}{\cos \xi_1 + \cosh(\psi_1 + \phi_1)}$$

$$= -2 \operatorname{arctanh} \frac{\sinh \psi_1 \sinh \phi_1}{\cos \xi_1 + \cosh \psi_1 \cosh \phi_1}, \tag{6.25}$$

considering (6.7) and (6.16)–(6.19).

We next consider the one-soliton solution of (6.4)–(6.6), which is simply derived from the one-periodic wave solution by taking the long-wave limit $k_1 \to 0$. To show this, we set

$$\xi_1^{(0)} = \pi \tag{6.26}$$

in (6.20) and take the limit $k_1 \to 0$, keeping a_1 finite. It follows from (6.20)–(6.22) that

$$\phi_1 \simeq k_1/(a_1 - 2\gamma), \tag{6.27}$$

$$\psi_1 \simeq k_1/a_1, \tag{6.28}$$

$$\cos \xi_1 \simeq -1 + (k_1^2/2)(x - a_1 t - x_{01}), \tag{6.29}$$

$$\sinh \phi_1 \simeq k_1/(a_1 - 2\gamma), \tag{6.30}$$

$$\sinh \psi_1 \simeq k_1/a_1, \tag{6.31}$$

$$\cosh \phi_1 \simeq 1 + \tfrac{1}{2}[k_1/(a_1 - 2\gamma)]^2, \tag{6.32}$$

$$\cosh \psi_1 \simeq 1 + \tfrac{1}{2}(k_1/a_1)^2. \tag{6.33}$$

Substituting these expressions into (6.25), we obtain, in the limit of $k_1 \to 0$, the one-soliton solution

$$w = \ln \frac{(x - a_1 t - x_{01})^2 + [1/a_1 - 1/(a_1 - 2\gamma)]^2}{(x - a_1 t - x_{01})^2 + [1/a_1 + 1/(a_1 - 2\gamma)]^2}$$

$$= -2 \operatorname{arctanh} \frac{2/a_1(a_1 - 2\gamma)}{(x - a_1 t - x_{01})^2 + [1/a_1 + 1/(a_1 - 2\gamma)]^2}. \tag{6.34}$$

Note that (6.34) behaves asymptotically for large x as

$$w \simeq -[4/a_1(a_1 - 2\gamma)]x^{-2}, \tag{6.35}$$

which has the same asymptotic form as that of the BO soliton solution [see (4.14)].

For general N, the solutions are constructed following the procedure developed in Section 2.2. The N-periodic wave solution may be expressed as [32]

$$f = \sum_{\mu=0,1} \exp\left[\sum_{j=1}^{N} \mu_j(i\xi_j - \phi_j + \psi_j) + \sum_{j<k}^{(N)} \mu_j \mu_k A_{jk}\right], \tag{6.36}$$

$$f' = \sum_{\mu=0,1} \exp\left[\sum_{j=1}^{N} \mu_j(i\xi_j - \phi_j - \psi_j) + \sum_{j<k}^{(N)} \mu_j \mu_k A_{jk}\right], \tag{6.37}$$

$$g = \sum_{\mu=0,1} \exp\left[\sum_{j=1}^{N} \mu_j(i\xi_j + \phi_j + \psi_j) + \sum_{j<k}^{(N)} \mu_j \mu_k A_{jk}\right], \tag{6.38}$$

$$g' = \sum_{\mu=0,1} \exp\left[\sum_{j=1}^{N} \mu_j(i\xi_j + \phi_j - \psi_j) + \sum_{j<k}^{(N)} \mu_j \mu_k A_{jk}\right], \tag{6.39}$$

where

$$\xi_j = k_j(x - a_j t - x_{0j}) + \xi_j^{(0)}, \qquad j = 1, 2, \ldots, N, \tag{6.40}$$

$$a_j = k_j \coth \phi_j + 2\gamma, \qquad j = 1, 2, \ldots, N, \tag{6.41}$$

$$\exp(2\psi_j) = (a_j + k_j)/(a_j - k_j), \qquad j = 1, 2, \ldots, N, \tag{6.42}$$

$$\exp A_{jl} = \frac{(a_j - a_l)^2 - (k_j - k_l)^2}{(a_j - a_l)^2 - (k_j + k_l)^2}, \qquad j, l = 1, 2, \ldots, N, \tag{6.43}$$

$$(\psi_j - \phi_j)/k_j > 0, \qquad j = 1, 2, \ldots, N, \tag{6.44}$$

$$(\psi_j + \phi_j)/k_j < 0, \qquad j = 1, 2, \ldots, N, \tag{6.45}$$

with k_j, ϕ_j, x_{0j}, and $\xi_j^{(0)}$ $(j = 1, 2, \ldots, N)$ being real constants.

The present N-periodic wave solution reduces to the N-soliton solution in the long-wave limit. The procedure to derive the N-soliton solution is the same as that developed for the BO N-soliton solution (see Section 3.1), that is, we set

$$\xi_j^{(0)} = \pi, \qquad j = 1, 2, \ldots, N, \tag{6.46}$$

and take the long-wave limit $k_j \to 0$ $(j = 1, 2, \ldots, N)$, keeping a_j $(j = 1, 2, \ldots, N)$ finite. The result is expressed in terms of the bilinear variables as

$$w = \ln(\tilde{g}'\tilde{f}/\tilde{g}\tilde{f}'). \tag{6.47}$$

Here

$$\tilde{f} = \det M, \tag{6.48}$$

$$\tilde{f}' = \det M', \tag{6.49}$$

$$\tilde{g} = \det L, \tag{6.50}$$

$$\tilde{g}' = \det L', \tag{6.51}$$

where M, M', L, and L' are $N \times N$ matrices whose elements are given by

$$M_{jk} = [i\theta_j - 1/(a_j - 2\gamma) + 1/a_j]\delta_{jk} + [2/(a_j - a_k)](1 - \delta_{jk}), \tag{6.52}$$

$$M'_{jk} = [i\theta_j - 1/(a_j - 2\gamma) - 1/a_j]\delta_{jk} + [2/(a_j - a_k)](1 - \delta_{jk}), \tag{6.53}$$

$$L_{jk} = [i\theta_j + 1/(a_j - 2\gamma) + 1/a_j]\delta_{jk} + [2/(a_j - a_k)](1 - \delta_{jk}), \tag{6.54}$$

$$L'_{jk} = [i\theta_j + 1/(a_j - 2\gamma) - 1/a_j]\delta_{jk} + [2/(a_j - a_k)](1 - \delta_{jk}), \tag{6.55}$$

with

$$\theta_j = x - a_j t - x_{0j}, \qquad j = 1, 2, \ldots, N. \tag{6.56}$$

The asymptotic forms of the present N-soliton solution for large values of t are readily obtained from (6.47)–(6.56) as

$$w \underset{t \to \pm\infty}{\simeq} \sum_{j=1}^{N} \ln \frac{\theta_j^2 + [1/a_j - 1/(a_j - 2\gamma)]^2}{\theta_j^2 + [1/a_j + 1/(a_j - 2\gamma)]^2}, \tag{6.57}$$

which implies that no phase shift occurs on collision of solitons. The situation is the same as that for the BO N-soliton solution discussed in Chapter 4.

From the Bäcklund transformation of the modified BO equation, a new nonlinear evolution equation, which may be called the second modified BO equation, can be generated, and this process may continue infinitely. The concept of the chain of the Bäcklund transformation for the KdV equation first appeared in [53]. As is well known, the modified KdV equation is generated from the Bäcklund transformation of the KdV equation [105]. Similarly, the second modified KdV equation is constructed from the Bäcklund transformation of the modified KdV equation [53]. Furthermore, the modified finite-depth fluid equation [81] and the modified Sine–Gordon equation [82] are also generated from the Bäcklund transformations of the

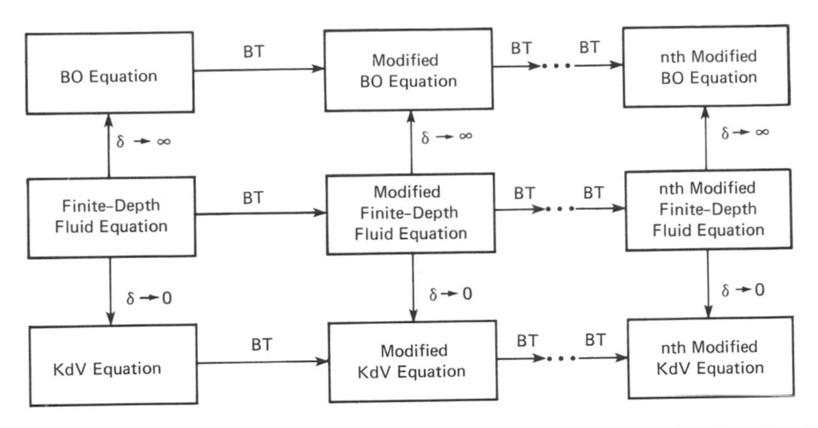

Fig. 6.1 New nonlinear evolution equations generated from the chain of Bäcklund transformations of the finite-depth fluid, the BO, and the KdV equations. Here BT denotes the Bäcklund transformation and δ the depth of fluid.

finite-depth fluid equation and the Sine–Gordon equation, respectively. In Fig. 6.1, the new nonlinear evolution equations generated by the chain of the Bäacklund transformations of the finite-depth fluid, the BO, and the KdV equations are depicted along with their interrelationships.

6.2 The Derivative Nonlinear Schrödinger Equation

In this section we shall consider the nonlinear evolution equation

$$ia_t + a_{xx} = ia(|a|^2)_x, \tag{6.58}$$

where a is the complex amplitude of the basic wave train. This equation governs the nonlinear self-modulation of weak periodic wave solutions of the BO equation [34]. It should be noted that Eq. (6.58) is different from the derivative nonlinear Schrödinger equation

$$ia_t + a_{xx} = i(a|a|^2)_x, \tag{6.59}$$

which is known to be completely integrable [118], and also different from the derivative nonlinear Schrödinger equation as expressed in (2.259).

We now seek a solution of (6.58) in the form [34]

$$a(x, t) = A(x)e^{-i\Omega t}, \tag{6.60}$$

$$A(x) = F(x)e^{i\phi(x)}, \tag{6.61}$$

where $F(x)$ and $\phi(x)$ are real functions and Ω is a constant. Substituting (6.60) and (6.61) into (6.58) yields a system of ordinary differential equations

$$(du/dx)^2 = \tfrac{1}{3}u^4 - 4(\Omega - c_1)u^2 + 4c_2 - 4c_1^2 \equiv V(u), \tag{6.62}$$

$$d\phi/dx = \tfrac{1}{2}u + c_1 u^{-1}, \tag{6.63}$$

where

$$u = F^2 \tag{6.64}$$

and c_1 and c_2 are integration constants. To obtain a bounded solution of (6.62), we decompose $V(u)$ in the form

$$V(u) = (u - \alpha)(u - \beta)(u - \gamma)(u - \delta), \tag{6.65}$$

with the condition

$$\delta \leq 0 \leq \gamma \leq \beta \leq \alpha. \tag{6.66}$$

An oscillating solution may then be expressed in terms of the Jacobian elliptic function (cn) as

$$u = \frac{\beta(\gamma - \delta) + \delta(\beta - \gamma)\,\mathrm{cn}^2(\xi, k)}{(\gamma - \delta) + (\beta - \gamma)\,\mathrm{cn}^2(\xi, k)}, \tag{6.67}$$

where

$$\xi = \sqrt{(\alpha - \gamma)(\beta - \delta)/12}\,x \tag{6.68}$$

and

$$k = \sqrt{\frac{(\alpha - \delta)(\beta - \gamma)}{(\alpha - \gamma)(\beta - \delta)}}. \tag{6.69}$$

A solitary wave solution is obtained from (6.67)–(6.69) by setting $\alpha = \beta$, for example, as

$$\begin{aligned} u &= \frac{2\beta(\beta + \gamma) - (2\beta + \gamma)(\beta - \gamma)\,\mathrm{sech}^2\,\xi}{2(\beta + \gamma) + (\beta - \gamma)\,\mathrm{sech}^2\,\xi} \\ &= \beta - \frac{(\beta - \gamma)(3\beta + 2\gamma)}{(\beta + \gamma)\cosh 2\xi + 2\beta}, \end{aligned} \tag{6.70}$$

with

$$\xi = \sqrt{(\beta - \gamma)(3\beta + \gamma)/12}\,x, \tag{6.71}$$

where use has been made of the relations

$$k = 1, \tag{6.72}$$

$$\mathrm{cn}(\xi, 1) = \mathrm{sech}\,\xi, \tag{6.73}$$

$$\alpha + \beta + \gamma + \delta = 0. \tag{6.74}$$

Since (6.58) is invariant under the transformations

$$\tilde{a}(\tilde{x}, \tilde{t}) = a(x, t) \exp[i(-\lambda x/2 + \lambda^2 t/4 + \phi_0)], \tag{6.75}$$

$$\tilde{x} = x - \lambda t - x_0, \tag{6.76}$$

$$\tilde{t} = t - t_0, \tag{6.77}$$

where λ, x_0, t_0, and ϕ_0 are arbitrary real constants, a solitary wave solution with a propagation velocity λ may be constructed from the present stationary solution (6.70) by employing (6.75)–(6.77). Whether (6.58) possesses multisoliton solutions is currently an unsolved problem. However, the result of the numerical solution involving a head-on collision of two solitary waves [34] strongly suggests that (6.58) is completely integrable, and it may therefore have an N-soliton solution, an infinite number of conservation laws, the Bäcklund transformation, etc., which are common properties of completely integrable nonlinear evolution equations.

Finally, we note that (6.58) exhibits at least three conservation laws, their explicit forms being written as

$$I_1 = \int_{-\infty}^{\infty} (a^*a - a_0^*a_0)\, dx, \tag{6.78}$$

$$I_2 = \int_{-\infty}^{\infty} [(a^*a)^2 - (a_0^*a_0)^2 + i(a^*a_x - aa_x^*)]\, dx, \tag{6.79}$$

$$I_3 = \int_{-\infty}^{\infty} [(a^*a)^3 - (a_0^*a_0)^3 - \tfrac{3}{2}i(a^2a^*a_x^* - a^{*2}aa_x) + 3a_x^*a_x]\, dx, \tag{6.80}$$

where a_0 is a boundary value of a, that is,

$$a_0 = a(\pm\infty, t). \tag{6.81}$$

6.3 The Perturbed Benjamin–Ono Equation

Throughout this text, we have been concerned only with nonlinear evolution equations that include the effects of nonlinearity and dispersion; these may be appropriate model equations that describe physical systems. However, there exist many situations in which another important effect, that of dissipation, cannot be neglected. In these cases, the equations must be modified to include the effect of dissipation. In this final section we shall consider the system that is described by the BO equation with a small dissipation effect [119].

The model equation may be written in the form

$$u_t + 4uu_x + Hu_{xx} = \varepsilon R[u], \tag{6.82}$$

where R is a functional of u representing the effect of dissipation and ε is a small positive parameter. Equation (6.82) may be called the perturbed BO equation. When $\varepsilon = 0$, (6.82) reduces to the BO equation. Although the exact method for solving (6.82) is not known, we can treat it by a perturbation method.§ To apply this method to (6.82), introduce the multiple time scales

$$t_j = \varepsilon^j t, \qquad j = 0, 1, 2, \ldots, \tag{6.83}$$

and expand u into an asymptotic series

$$u = \sum_{j=0}^{\infty} \varepsilon^j u_j, \tag{6.84}$$

where u_j is assumed to be a function of $x, t_0, t_1, t_2, \ldots,$

$$u_j = u_j(x, t_0, t_1, t_2, \ldots), \qquad j = 0, 1, 2, \ldots. \tag{6.85}$$

As a consequence of (6.83), the time derivative is replaced by

$$\partial/\partial t = \sum_{j=0}^{\infty} \varepsilon^j \, \partial/\partial t_j. \tag{6.86}$$

Substituting (6.84)–(6.86) into (6.82) and comparing the ε^j $(j = 1, 2, \ldots)$ terms on both sides of (6.82), we obtain a system of equations, the first two of which are

$$O(\varepsilon^0): u_{0,t_0} + 4u_0 u_{0,x} + Hu_{0,xx} = 0, \tag{6.87}$$

$$O(\varepsilon^1): u_{1,t_0} + 4(u_0 u_1)_x + Hu_{1,xx} = R[u_0] - u_{0,t_1}. \tag{6.88}$$

§ The method used here is the multiple time-scale expansion [120, 121].

Note that the equations that derive from the ε^j terms are linear with respect to u_j for $j \geq 1$, while only the lowest-order equation (6.87) is nonlinear.

To illustrate the method, take the one-soliton solution of (6.87) as the lowest-order one

$$u_0 = a/(z^2 + 1), \tag{6.89}$$

with

$$z = a(x - \xi), \tag{6.90}$$

$$\xi = at_0 + \tilde{\xi}_0(t_1, t_2, \ldots), \qquad \tilde{\xi}_0(0, 0, \ldots) = \xi_0. \tag{6.91}$$

Amplitude a and phase $\tilde{\xi}_0$ are not constants but functions of $t_1, t_2, \ldots$. In the following discussion, we consider the expansion up to the order ε, so that the time variables $t_2, t_3, \ldots$ will not be written explicitly. The first-order equation (6.88) may then be solved, provided that the t_1 dependence of the lowest-order solution u_0 is found. To show this, consider the linear equation for f

$$L[f] \equiv f_{t_0} + 4(u_0 f)_x + Hf_{xx} = 0. \tag{6.92}$$

Function g defined by

$$g(x) = \int_{-\infty}^{x} f(x')\, dx', \tag{6.93}$$

satisfies the equation

$$L^*[g] \equiv g_{t_0} + 4u_0 g_x + Hg_{xx} = 0. \tag{6.94}$$

Operator L^* is an adjoint to L. Multiplying (6.88) by g and integrating with respect to x from $-\infty$ to ∞, we obtain

$$\left(\int_{-\infty}^{\infty} gu_1\, dx\right)_{t_0} = \int_{-\infty}^{\infty} g\{R[u_0] - u_{0,t_1}\}\, dx, \tag{6.95}$$

with the aid of (6.94). If g depends on x and t only in the combination $x - at_0$, then the right-hand side of (6.95) is not made to depend on t_0 by introducing a new integration variable $x - at_0$ instead of x. In this situation, (6.95) is integrated as

$$\int_{-\infty}^{\infty} gu_1\, dx \propto t_0, \tag{6.96}$$

which implies that u_1 diverges for large t_0. To eliminate this undesirable behavior, the following *nonsecularity condition* may be imposed:

$$\int_{-\infty}^{\infty} g\{R[u_0] - u_{0,t_1}\}\, dx = 0. \tag{6.97}$$

Functions f and g may easily be found as follows: Differentiate (6.87) with respect to the independent parameters p_j included in u_0. In the one-soliton case, we can set $p_1 = a$ and $p_2 = \xi$. Then

$$(\partial u_0/\partial p_j)_{t_0} + 4(u_0\, \partial u_0/\partial p_j)_x + H(\partial u_0/\partial p_j)_{xx} = 0, \tag{6.98}$$

which means that

$$L[\partial u_0/\partial p_j] = 0. \tag{6.99}$$

Therefore

$$f_j = \partial u_0/\partial p_j \tag{6.100}$$

and

$$g_j = \int_{-\infty}^{x} f_j\, dx'. \tag{6.101}$$

For the one-soliton case, we find from (6.89), (6.100), and (6.101) that

$$f_1 = \frac{\partial u_0}{\partial a} = \frac{1 - z^2}{(z^2 + 1)^2}, \tag{6.102}$$

$$g_1 = \int_{-\infty}^{x} \frac{\partial u_0}{\partial a}\, dx' = \frac{1}{a}\frac{z}{z^2 + 1}, \tag{6.103}$$

$$f_2 = \frac{\partial u_0}{\partial \xi} = \frac{2a^2 z}{(z^2 + 1)^2}, \tag{6.104}$$

$$g_2 = \int_{-\infty}^{x} \frac{\partial u_0}{\partial \xi}\, dx' = -\frac{a}{z^2 + 1}. \tag{6.105}$$

The right-hand side of (6.88) becomes

$$\begin{aligned} R[u_0] - u_{0,t_1} &= R[u_0] - \frac{\partial u_0}{\partial a}\frac{\partial a}{\partial t_1} - \frac{\partial u_0}{\partial \xi}\frac{\partial \xi}{\partial t_1} \\ &= R[u_0] - f_1 \frac{\partial a}{\partial t_1} - f_2 \frac{\partial \xi}{\partial t_1}, \end{aligned} \tag{6.106}$$

by the definition of f_1 and f_2. The nonsecularity conditions (6.97)

$$\int_{-\infty}^{\infty} g_j\{R[u_0] - u_{0,t_1}\}\, dx = 0, \qquad j = 1, 2, \tag{6.107}$$

now yield equations for a and ξ as

$$\frac{\partial a}{\partial t_1} = \frac{4}{\pi}\int_{-\infty}^{\infty} \frac{R[u_0(z)]}{z^2 + 1}\, dz, \tag{6.108}$$

$$\frac{\partial \xi}{\partial t_1} = \frac{4}{\pi}\frac{1}{a^2}\int_{-\infty}^{\infty} \frac{z}{z^2 + 1} R[u_0(z)]\, dz. \tag{6.109}$$

These results may be rewritten in terms of the original variable t as

$$\frac{da}{dt} = \frac{4\varepsilon}{\pi}\int_{-\infty}^{\infty} \frac{R[u_0(z)]}{z^2 + 1}\, dz, \tag{6.110}$$

$$\frac{d\xi}{dt} = a + \frac{4\varepsilon}{\pi}\frac{1}{a^2}\int_{-\infty}^{\infty} \frac{z}{z^2 + 1} R[u_0(z)]\, dz, \tag{6.111}$$

by noting

$$d\xi/dt = \partial\xi/\partial t_0 + \varepsilon\, \partial\xi/\partial t_1 + O(\varepsilon^2) \tag{6.112}$$

and (6.91).

This result can be extended to the case where the lowest-order solution u_0 consists of N solitons, and the independent parameters included in u_0 are N amplitudes a_j ($j = 1, 2, \ldots, N$) and N phases ξ_j ($j = 1, 2, \ldots, N$). If we set these parameters as

$$p_j = a_j, \qquad j = 1, 2, \ldots, N, \tag{6.113}$$

$$p_{N+j} = \xi_j, \qquad j = 1, 2, \ldots, N, \tag{6.114}$$

then the $2N$ independent solutions of (6.94) are constructed as

$$g_j(x) = \int_{-\infty}^{x} \frac{\partial u_0}{\partial p_j}\, dx', \qquad j = 1, 2, \ldots, 2N. \tag{6.115}$$

The nonsecularity conditions (6.97) are then replaced by $2N$ simultaneous equations for $2N$ unknown quantities $\partial p_j/\partial t_1$ ($j = 1, 2, \ldots, 2N$)

$$\sum_{s=1}^{2N} \frac{\partial p_s}{\partial t_1}\left(\frac{\partial u_0}{\partial p_s}, g_j\right) = \{R[u_0], g_j\}, \qquad j = 1, 2, \ldots, 2N, \tag{6.116}$$

where

$$(f, g) = \int_{-\infty}^{\infty} fg\, dx, \tag{6.117}$$

from which we can determine the t_1 dependences of the parameters p_j ($j = 1, 2, \ldots, 2N$).

As explicit examples, we shall consider two forms of R given as

$$R[u] = -u, \tag{6.118}$$

$$R[u] = u_{xx}. \tag{6.119}$$

For the one-soliton case, we obtain from (6.110) and (6.111) the following results:

(i) $R[u] = -u$

$$a = a_0 e^{-2\varepsilon t}, \tag{6.120}$$

$$\xi = \xi_0 + (a_0/2\varepsilon)(1 - e^{-2\varepsilon t}). \tag{6.121}$$

(ii) $R[u] = u_{xx}$

$$a = a_0(1 + 2\varepsilon a_0^2 t)^{-1/2}, \tag{6.122}$$

$$\xi = \xi_0 + (1/\varepsilon a_0)[(1 + 2\varepsilon a_0^2 t)^{1/2} - 1], \tag{6.123}$$

where a_0 and ξ_0 are initial amplitude and initial phase of a soliton, respectively.

Obviously, these results reduce to the unperturbed cases when $\varepsilon = 0$. Once the t_1 dependences of parameters have been determined, one can proceed to the first-order equation (6.88), which describes the time evolution of the nonsoliton part. Being a linear equation for u_1, it may be more tractable than the original nonlinear equation (6.82). However, the details will not be discussed here.

Finally, we shall comment on the conservation laws. Because of the effect of dissipation, the terms I_n given by (3.187)–(3.191) are no longer conserved and depend explicitly on time. The time evolutions of I_n can be expressed as

$$\frac{dI_n}{dt} = \varepsilon \int_{-\infty}^{\infty} \frac{\delta I_n}{\delta u} R[u]\, dx, \qquad n = 1, 2, \ldots. \tag{6.124}$$

For the one-soliton case, the approximate solution of (6.82)

$$u = u_0 + \varepsilon u_1 \tag{6.125}$$

satisfies (6.124) up to the order ε. To show this, note that

$$I_n = \pi(a/4)^{n-1}, \qquad n = 1, 2, \ldots, \tag{6.126}$$

for the one-soliton solution (6.89) [see (3.256)]. From (6.126), we obtain

$$\delta I_n/\delta u = (\pi/4)(n - 1)(a/4)^{n-2}\,\delta a/\delta u, \qquad n = 2, 3, \ldots. \tag{6.127}$$

On the other hand, it follows from (6.126) with $n = 2$ that

$$\tfrac{1}{2}\int_{-\infty}^{\infty} u^2\,dx = (\pi/4)a, \tag{6.128}$$

which, by taking the functional derivative, yields

$$u = (\pi/4)\,\delta a/\delta u. \tag{6.129}$$

Combining (6.127) with (6.129), we obtain the useful relations

$$\delta I_1/\delta u = 1, \tag{6.130}$$

$$\delta I_n/\delta u = (n - 1)(a/4)^{n-2}u = 4(n - 1)(a/4)^{n-1}[1/(z^2 + 1)], \qquad n \geq 2. \tag{6.131}$$

Then

$$\begin{aligned}
\frac{dI_n}{dt} &= \sum_{j=0}^{\infty} \varepsilon^j \frac{\partial}{\partial t_j} I_n(u_0 + \varepsilon u_1) \\
&= \varepsilon\left[\frac{\partial I_n(u_0)}{\partial t_1} + \frac{\partial}{\partial t_0}\int_{-\infty}^{\infty} \frac{\delta I_n}{\delta u}\bigg|_{u=u_0} u_1\,dx\right] + O(\varepsilon^2) \\
&= \varepsilon\left[\int_{-\infty}^{\infty} \frac{\delta I_n}{\delta u}\bigg|_{u=u_0}\left(\frac{\partial u_0}{\partial t_1} + \frac{\partial u_1}{\partial t_0}\right)\right. \\
&\qquad \left. + \int_{-\infty}^{\infty}\left(\frac{\partial}{\partial t_0}\frac{\delta I_n}{\delta u}\bigg|_{u=u_0}\right)u_1\right] dx + O(\varepsilon^2),
\end{aligned} \tag{6.132}$$

in view of the conservation laws

$$\partial I_n(u_0)/\partial t_0 = 0, \qquad n = 1, 2, \ldots. \tag{6.133}$$

Substituting (6.88) into (6.132) and integrating by parts, we obtain

$$\frac{dI_n}{dt} = \varepsilon \int_{-\infty}^{\infty} \frac{\delta I_n}{\delta u}\bigg|_{u=u_0} R[u_0]\, dx + \varepsilon \int_{-\infty}^{\infty} u_1 \left(\frac{\partial}{\partial t_0} + 4u_0 \frac{\partial}{\partial x} + H \frac{\partial^2}{\partial x^2}\right) \frac{\delta I_n}{\delta u}\bigg|_{u=u_0} dx + O(\varepsilon^2). \tag{6.134}$$

However, the second term on the right-hand side of (6.134) vanishes because of (6.131), that is,

$$\delta I_n/\delta u|_{u=u_0} \propto u_0, \tag{6.135}$$

and (6.87), the lowest-order equation for u_0. Thus,

$$\frac{dI_n}{dt} = \varepsilon \int_{-\infty}^{\infty} \frac{\delta I_n}{\delta u}\bigg|_{u=u_0} R[u_0]\, dx + O(\varepsilon^2). \tag{6.136}$$

On the other hand, it follows from (6.124) that

$$\varepsilon \int_{-\infty}^{\infty} \frac{\delta I_n}{\delta u} R[u]\, dx = \varepsilon \int_{-\infty}^{\infty} \frac{\delta I_n}{\delta u}\bigg|_{u=u_0} R[u_0]\, dx + O(\varepsilon^2). \tag{6.137}$$

Therefore, we see from (6.136) and (6.137) that relation (6.124) is satisfied by (6.125) up to the order ε.

The multiple time-scale expansion method developed here can be applied to a wide class of nonlinear dispersive and dissipative equations. The interested reader may refer to [122, 123, 124].

Appendix I

Formulas of the Bilinear Operators[§]

The definition of the bilinear operators is given by

$$D_t^n D_x^m a \cdot b = (\partial/\partial t - \partial/\partial t')^n (\partial/\partial x - \partial/\partial x')^m a(x, t) b(x', t')|_{\substack{x'=x \\ t'=t}}.$$

1. The following formulas are easily derived from the definition of the bilinear operators:

$$D_x^m a \cdot 1 = \partial^m a/\partial x^m, \tag{App.I.1.1}$$

$$D_x^m a \cdot b = (-1)^m D_x^m b \cdot a, \tag{App.I.1.2}$$

$$D_x^m a \cdot a = 0 \qquad \text{for} \quad \text{odd } m, \tag{App.I.1.3}$$

$$D_x D_t a \cdot 1 = D_x D_t 1 \cdot a = \partial^2 a/\partial x \partial t, \tag{App.I.1.4}$$

$$D_x a \cdot b = 0 \leftrightarrow a \propto b, \tag{App.I.1.5}$$

$$D_x^{2m+1} a_t \cdot b = \tfrac{1}{2} D_x^{2m+1} D_t a \cdot b. \tag{App.I.1.6}$$

[§] The formulas given in this Appendix are found mainly in References [5], [10], and [13].

2. The following formula is a fundamental property of the bilinear operator:

$$\exp(\varepsilon D_x)a(x)\cdot b(x) = a(x+\varepsilon)b(x-\varepsilon). \qquad \text{(App.I.2)}$$

This formula is proved as follows:

$$\begin{aligned}
&\exp(\varepsilon D_x)a(x)\cdot b(x)\\
&\quad = \sum_{n=0}^{\infty}\frac{\varepsilon^n}{n!}\left(\frac{\partial}{\partial x}-\frac{\partial}{\partial x'}\right)^n a(x)b(x')\bigg|_{x'=x} && \text{by definition}\\
&\quad = \sum_{n=0}^{\infty}\frac{\varepsilon^n}{n!}\sum_{s=0}^{n}(-1)^{n-s}{}_nC_s\frac{\partial^s a}{\partial x^s}\frac{\partial^{n-s}b}{\partial x^{n-s}}\\
&\quad = \sum_{s=0}^{\infty}\sum_{n=s}^{\infty}\frac{(-1)^{n-s}\varepsilon^n}{s!(n-s)!}\frac{\partial^s a}{\partial x^s}\frac{\partial^{n-s}b}{\partial x^{n-s}} && \text{by exchanging the order of sum}\\
&\quad = \sum_{s=0}^{\infty}\frac{\varepsilon^s}{s!}\frac{\partial^s a}{\partial x^s}\sum_{n=0}^{\infty}\frac{(-\varepsilon)^n}{n!}\frac{\partial^n b}{\partial x^n} && \text{by replacing } n \text{ by } n+s\\
&\quad = a(x+\varepsilon)b(x-\varepsilon).
\end{aligned}$$

If we set $a(x) = \exp(p_1 x)$ and $b(x) = \exp(p_2 x)$ in (App.I.2) and compare the ε^m term on both sides of (App.I.2), we obtain the formula

$$D_x^m \exp(p_1 x)\cdot\exp(p_2 x) = (p_1 - p_2)^m \exp(p_1 + p_2)x. \qquad \text{(App.I.2.1)}$$

Let $F(D_t, D_x)$ be a polynomial of D_t and D_x. Then it follows from (App.I.1.1) and (App.I.2.1) that

$$\begin{aligned}
&F(D_t, D_x)\exp(\Omega_1 t + p_1 x)\cdot\exp(\Omega_2 t + p_2 x)\\
&\quad = \frac{F(\Omega_1 - \Omega_2, p_1 - p_2)}{F(\Omega_1 + \Omega_2, p_1 + p_2)}F(D_t, D_x)\\
&\quad\quad \times \exp[(\Omega_1 + \Omega_2)t + (p_1 + p_2)x]\cdot 1, \qquad \text{(App.I.2.2)}
\end{aligned}$$

which is very useful in deriving the N-soliton solution.

3. The following formulas are useful in transforming bilinear equations into the original forms of nonlinear equations:

$$\begin{aligned}
\exp(\varepsilon D_x + \delta D_t)a\cdot b = \exp[&\sinh(\varepsilon\,\partial/\partial x + \delta\,\partial/\partial t)\ln(a/b)\\
&+ \cosh(\varepsilon\,\partial/\partial x + \delta\,\partial/\partial t)\ln(ab)], \qquad \text{(App.I.3)}
\end{aligned}$$

where $a = a(x, t)$ and $b = b(x, t)$. The left-hand side of (App.I.3) is $a(x + \varepsilon, t + \delta)b(x - \varepsilon, t - \delta)$ by (App.I.2). On the other hand, the right-hand side of (App.I.3) reduces to

$$\begin{aligned}\exp\{&\tfrac{1}{2} \ln[a(x + \varepsilon, t + \delta)/b(x + \varepsilon, t + \delta)] \\ &- \tfrac{1}{2} \ln[a(x - \varepsilon, t - \delta)/b(x - \varepsilon, t - \delta)] \\ &+ \tfrac{1}{2} \ln[a(x + \varepsilon, t + \delta)b(x + \varepsilon, t + \delta)] \\ &+ \tfrac{1}{2} \ln[a(x - \varepsilon, t - \delta)b(x - \varepsilon, t - \delta)]\} \\ &\quad = \exp\{\ln[a(x + \varepsilon, t + \delta)b(x - \varepsilon, t - \delta)]\} \\ &\quad = a(x + \varepsilon, t + \delta)b(x - \varepsilon, t - \delta),\end{aligned}$$

where we have used Taylor's formula

$$\exp(\varepsilon\, \partial/\partial x + \delta\, \partial/\partial t)a(x, t) = a(x + \varepsilon, t + \delta).$$

Therefore, (App.I.3) has been proved.

Let $\phi = \ln(a/b)$ and $\rho = \ln(ab)$. It follows from (App.I.3) with $\delta = 0$, expanding both sides in powers of ε, and comparing the ε^n terms, that

$$(D_x a \cdot b)/ab = \phi_x, \qquad \text{(App.I.3.1)}$$

$$(D_x^2 a \cdot b)/ab = \rho_{xx} + (\phi_x)^2, \qquad \text{(App.I.3.2)}$$

$$(D_x^3 a \cdot b)/ab = \phi_{xxx} + 3\phi_x \rho_{xx} + (\phi_x)^3, \qquad \text{(App.I.3.3)}$$

$$(D_x D_t a \cdot b)/ab = \rho_{xt} + \phi_{xt}. \qquad \text{(App.I.3.4)}$$

4. The following formula is a special case of (App.I.3):

$$\exp(\varepsilon D_x)a \cdot b = \{\exp[2 \cosh(\varepsilon\, \partial/\partial x) \ln b]\}[\exp(\varepsilon\, \partial/\partial x)(a/b)], \qquad \text{(App.I.4)}$$

which follows by setting $\delta = 0$ and noting the relation

$$\begin{aligned}\exp[\exp(\varepsilon\, \partial/\partial x) \ln(a/b)] &= \exp\{\ln[a(x + \varepsilon)/b(x + \varepsilon)]\} \\ &= a(x + \varepsilon)/b(x + \varepsilon) = \exp(\varepsilon\, \partial/\partial x)(a/b).\end{aligned}$$

5. If we set $a = b = f$ in (App.I.4) and use (App.I.1.3), we have

$$\cosh(\varepsilon D_x) f \cdot f = \exp[2 \cosh(\varepsilon\, \partial/\partial x) \ln f]. \qquad \text{(App.I.5)}$$

Let $u = 2(\ln f)_{xx}$ and $u_{nx} = \partial^n u/\partial x^n$ ($n = 0, 1, 2, \ldots$). Comparing the ε^n terms on both sides of (App.I.5) yields the formulas

$$(D_x^2 f \cdot f)/f^2 = u, \tag{App.I.5.1}$$

$$(D_x^4 f \cdot f)/f^2 = u_{2x} + 3u^2, \tag{App.I.5.2}$$

$$(D_x^6 f \cdot f)/f^2 = u_{4x} + 15uu_{2x} + 15u^3, \tag{App.I.5.3}$$

$$(D_x^8 f \cdot f)/f^2 = u_{6x} + 28uu_{4x} + 35(u_{2x})^2 + 210u^2u_{2x} + 105u^4, \tag{App.I.5.4}$$

$$\begin{aligned}(D_x^{10} f \cdot f)/f^2 = u_{8x} &+ 45uu_{6x} + 210u_{2x}u_{4x} + 1575u(u_{2x})^2 \\ &+ 630u^2u_{4x} + 3150u^3u_{2x} + 945u^5.\end{aligned} \tag{App.I.5.5}$$

6. The following formula is very important in discussing the Bäcklund transformation in bilinear formalism:

$$\begin{aligned}&\exp(D_1)[\exp(D_2)a \cdot b] \cdot [\exp(D_3)c \cdot d] \\ &\quad = \exp \tfrac{1}{2}(D_2 - D_3)\{\exp[\tfrac{1}{2}(D_2 + D_3) + D_1]a \cdot d\} \\ &\qquad \cdot \{\exp[\tfrac{1}{2}(D_2 + D_3) - D_1]c \cdot b\},\end{aligned} \tag{App.I.6}$$

where

$$D_j = \varepsilon_j D_x + \delta_j D_t, \qquad j = 1, 2, 3,$$

with ε_j and δ_j being arbitrary constants. Using (App.I.2), (App.I.6) is proved as follows: The left-hand side of (App.I.6) becomes

$$\begin{aligned}&\exp(D_1)[a(x + \varepsilon_2, t + \delta_2)b(x - \varepsilon_2, t - \delta_2)] \\ &\qquad \cdot [c(x + \varepsilon_3, t + \delta_3)d(x - \varepsilon_3, t - \delta_3)] \\ &\quad = a(x + \varepsilon_1 + \varepsilon_2, t + \delta_1 + \delta_2)b(x + \varepsilon_1 - \varepsilon_2, t + \delta_1 - \delta_2) \\ &\qquad \times c(x - \varepsilon_1 + \varepsilon_3, t - \delta_1 + \delta_3)d(x - \varepsilon_1 - \varepsilon_3, t - \delta_1 - \delta_3).\end{aligned}$$

On the other hand, the right-hand side of (App.I.6) reduces to

$$\begin{aligned}&\exp[\tfrac{1}{2}(D_2 - D_3)][a(x + \varepsilon_1 + (\varepsilon_2 + \varepsilon_3)/2, t + \delta_1 + (\delta_2 + \delta_3)/2) \\ &\qquad \times d(x - \varepsilon_1 - (\varepsilon_2 + \varepsilon_3)/2, t - \delta_1 - (\delta_2 + \delta_3)/2)] \\ &\qquad \cdot [c(x - \varepsilon_1 + (\varepsilon_2 + \varepsilon_3)/2, t - \delta_1 + (\delta_2 + \delta_3)/2) \\ &\qquad \times b(x + \varepsilon_1 - (\varepsilon_2 + \varepsilon_3)/2, t + \delta_1 - (\delta_2 + \delta_3)/2)] \\ &\quad = a(x + \varepsilon_1 + \varepsilon_2, t + \delta_1 + \delta_2)d(x - \varepsilon_1 - \varepsilon_3, t - \delta_1 - \delta_3) \\ &\qquad \times c(x - \varepsilon_1 + \varepsilon_3, t - \delta_1 + \delta_3)b(x + \varepsilon_1 - \varepsilon_2, t + \delta_1 - \delta_2),\end{aligned}$$

which implies (App.I.6). The following formulas are derived from (App.I.6) by appropriate choice of ε_j and δ_j:

$$(D_x a \cdot b)cd - ab(D_x c \cdot d) = (D_x a \cdot c)bd - ac(D_x b \cdot d) = D_x ad \cdot bc, \tag{App.I.6.1}$$

$$\begin{aligned} &D_x[(D_t a \cdot b) \cdot cd + ab \cdot (D_t c \cdot d)] \\ &\quad = (D_t D_x a \cdot d)cb - ad(D_t D_x c \cdot b) + (D_x a \cdot d)(D_t c \cdot b) \\ &\qquad - (D_t a \cdot d)(D_x c \cdot b), \end{aligned} \tag{App.I.6.2}$$

$$(D_x^2 a \cdot b)cd - ab(D_x^2 c \cdot d) = D_x[(D_x a \cdot d) \cdot cb + ad \cdot (D_x c \cdot b)], \tag{App.I.6.3}$$

$$\begin{aligned} &(D_x^3 a \cdot b)cd - ab(D_x^3 c \cdot d) \\ &\quad = \tfrac{1}{4} D_x^3 ad \cdot cb + \tfrac{3}{4} D_x[(D_x^2 a \cdot d) \cdot cb \\ &\qquad + 2(D_x a \cdot d) \cdot (D_x c \cdot b) + ad \cdot (D_x^2 c \cdot b)], \end{aligned} \tag{App.I.6.4}$$

$$\begin{aligned} &(D_x^4 a \cdot a)cc - aa(D_x^4 c \cdot c) \\ &\quad = 2D_x(D_x^3 a \cdot c) \cdot ca + 6D_x(D_x^2 a \cdot c) \cdot (D_x c \cdot a) \\ &\quad = 2D_x^3(D_x a \cdot c) \cdot ac. \end{aligned} \tag{App.I.6.5}$$

7. The following formula is also useful in transforming the Bäcklund transformation written in bilinear operators into a form written in original variables:

$$\begin{aligned} &[\exp(\delta D_x)a \cdot b]cd + ab[\exp(\delta D_x)c \cdot d] \\ &\quad = [\exp(\delta D_x)a \cdot d]cb + ad[\exp(\delta D_x)c \cdot b] \\ &\qquad - \exp(\delta D_x/2)[2 \sinh(\delta D_x/2)a \cdot c] \cdot [2 \sinh(\delta D_x/2)b \cdot d], \end{aligned} \tag{App.I.7}$$

which may be verified by repeated use of (App.I.2). From (App.I.7), we obtain the formulas

$$(D_x a \cdot b)cd + ab(D_x c \cdot d) = (D_x a \cdot d)cb + ad(D_x c \cdot b), \tag{App.I.7.1}$$

$$\begin{aligned} &(D_x^2 a \cdot d)cb + ad(D_x^2 c \cdot b) \\ &\quad = (D_x^2 a \cdot d)cb + ad(D_x^2 c \cdot b) - 2(D_x a \cdot c)(D_x b \cdot d). \end{aligned} \tag{App.I.7.2}$$

8. To prove the commutability relation, the following formula is utilized:

$$\begin{aligned} &[\exp(\delta\, \partial/\partial x) \exp(\varepsilon D_x)a \cdot b]c - [\exp(\delta\, \partial/\partial x) \exp(\varepsilon D_x)a \cdot c]b \\ &\quad = 2\{\exp[(\varepsilon + \delta)\, \partial/\partial x]a\} \\ &\qquad \times \{\exp[\tfrac{1}{2}(-\varepsilon + \delta)\, \partial/\partial x] \sinh[\tfrac{1}{2}(-\varepsilon + \delta)D_x]b \cdot c\}. \end{aligned} \tag{App.I.8}$$

This formula is proved by using (App.I.2). It follows by comparing the terms $\delta^m \varepsilon^n$ ($m, n = 0, 1, 2, \ldots$) on both sides of (App.I.8) that

$$(D_x a \cdot b)c - (D_x a \cdot c)b = -a(D_x b \cdot c), \tag{App.I.8.1}$$

$$(D_x^2 a \cdot b)c - (D_x^2 a \cdot c)b = -2a_x(D_x b \cdot c) + a(D_x b \cdot c)_x, \tag{App.I.8.2}$$

$$\begin{aligned}(D_x^3 a \cdot b)c - (D_x^3 a \cdot c)b = &-3a_{xx}(D_x b \cdot c) + 3a_x(D_x b \cdot c)_x \\ &- \tfrac{1}{4}a[(D_x^3 b \cdot c) + 3(D_x b \cdot c)_{xx}],\end{aligned} \tag{App.I.8.3}$$

$$\begin{aligned}(D_x^2 a \cdot b)_x c - (D_x^2 a \cdot c)_x b = &-a_{xx}(D_x b \cdot c) - a_x(D_x b \cdot c)_x \\ &+ \tfrac{1}{4}a[(D_x^3 b \cdot c) + 3(D_x b \cdot c)_{xx}].\end{aligned} \tag{App.I.8.4}$$

Appendix II

Properties of the Matrices M and A

We consider the $N \times N$ matrices whose elements are given by

$$M_{jk} = \delta_{jk}(i\theta_j + a_j^{-1}) + 2(1 - \delta_{jk})(a_j - a_k)^{-1}, \quad \text{(App.II.1)}$$

$$A_{jk} = \delta_{jk}(z + y_j^{-1}) + 2(1 - \delta_{jk})(y_j - y_k)^{-1}. \quad \text{(App.II.2)}$$

[See (3.17), (3.38), and (3.302).] These matrices have the properties

$$I \equiv \sum_{\substack{j,k=1 \\ (j \neq k)}}^{N} (M^s)_{jk} \left(\sum_{m=1}^{n} a_j^{n-m} a_k^{m-1} \right) = 0, \qquad s, n = 1, 2, \ldots, \quad \text{(App.II.3)}$$

$$\det A = (z + 1)^N. \quad \text{(App.II.4)}$$

We shall first prove (App.II.3). Using the definition of Kronecker's delta and the matrix product rule, I is converted to

$$I = \sum_{j,k=1}^{N} (M^s)_{jk}(1 - \delta_{jk})\left(\sum_{m=1}^{n} a_j^{n-m} a_k^{m-1}\right)$$

$$= \sum_{j_1,\ldots,j_{s+1}=1}^{N} M_{j_1j_2} M_{j_2j_3} \cdots M_{j_sj_{s+1}}(1 - \delta_{j_1j_{s+1}})\left(\sum_{m=1}^{n} a_{j_1}^{n-m} a_{j_{s+1}}^{m-1}\right)$$

$$= \sum_{j_1,\ldots,j_{s+1}=1}^{N} M_{j_1j_2} M_{j_2j_3} \cdots M_{j_sj_{s+1}} M_{j_{s+1}j_1}(M_{j_{s+1}j_1})^{-1}$$

$$\times (1 - \delta_{j_1j_{s+1}})\left(\sum_{m=1}^{n} a_{j_1}^{n-m} a_{j_{s+1}}^{m-1}\right). \tag{App.II.5}$$

On the other hand, it follows from the definition of the matrix M that

$$(M_{j_{s+1}j_1})^{-1} = (i\theta_{j_1} + a_{j_1}^{-1})^{-1}\delta_{j_{s+1}j_1} + \tfrac{1}{2}(a_{j_{s+1}} - a_j)(1 - \delta_{j_{s+1}j_1}). \tag{App.II.6}$$

Substituting (App.II.6) into (App.II.5) and using the properties of Kronecker's delta,

$$\delta_{jk}(1 - \delta_{jk}) = 0, \tag{App.II.7}$$

$$(1 - \delta_{jk})^2 = 1 - \delta_{jk}, \tag{App.II.8}$$

$$(a_j - a_k)\delta_{jk} = 0, \tag{App.II.9}$$

(App.II.5) reduces to

$$I = \sum_{j_1,\ldots,j_{s+1}=1}^{N} M_{j_1j_2} M_{j_2j_3} \cdots M_{j_{s+1}j_1}(a_{j_{s+1}} - a_{j_1})\left(\sum_{m=1}^{n} a_{j_1}^{n-m} a_{j_{s+1}}^{m-1}\right)\Big/2$$

$$= \sum_{j_1,\ldots,j_{s+1}=1}^{N} M_{j_1j_2} M_{j_2j_3} \cdots M_{j_{s+1}j_1}(a_{j_{s+1}}^n - a_{j_1}^n)/2, \tag{App.II.10}$$

where use has been made of the formula

$$(a_{j_{s+1}} - a_{j_1})\left(\sum_{m=1}^{n} a_{j_1}^{n-m} a_{j_{s+1}}^{m-1}\right) = a_{j_{s+1}}^n - a_{j_1}^n. \tag{App.II.11}$$

Consider the $s + 1$ cyclic permutations of the indices $j_1, j_2, \ldots, j_{s+1}$

$$(j_1, j_2, \ldots, j_{s+1}) \to (j_2, j_3, \ldots, j_1) \to (j_3, j_4, \ldots, j_2) \to \cdots$$
$$\to (j_{s+1}, j_1, \ldots, j_s). \tag{App.II.12}$$

Since the quantity $M_{j_1 j_2} M_{j_2 j_3} \cdots M_{j_{s+1} j_1}$ in (App.II.10) is invariant under these permutations, we obtain

$$I = 1/(s+1) \sum_{j_1, \ldots, j_{s+1}=1}^{N} M_{j_1 j_2} M_{j_2 j_3} \cdots M_{j_{s+1} j_1} \times [(a_{j_{s+1}}^n - a_{j_1}^n) + (a_{j_1}^n - a_{j_2}^n) + \cdots + (a_{j_s}^n - a_{j_{s+1}}^n)]/2 = 0, \quad \text{(App.II.13)}$$

which proves (App.II.3).

For $n = 1, 2$, (App.II.3) become

$$\sum_{\substack{j,k=1 \\ (j \neq k)}}^{N} (M^s)_{jk} = 0, \quad \text{(App.II.14)}$$

$$\sum_{\substack{j,k=1 \\ (j \neq k)}}^{N} (M^s)_{jk}(a_j + a_k) = 0, \quad \text{(App.II.15)}$$

respectively.

We now introduce the characteristic polynomial of the matrix M:

$$\det(\lambda I - M) = \lambda^N + c_1 \lambda^{N-1} + \cdots + c_{N-1}\lambda + c_N. \quad \text{(App.II.16)}$$

It follows from the well-known Hamilton–Cayley theorem that

$$M^N + c_1 M^{N-1} + \cdots + c_{N-1} M + c_N I = 0. \quad \text{(App.II.17)}$$

From (App.II.16) with $\lambda = 0$

$$c_N = (-1)^N \det M, \quad \text{(App.II.18)}$$

which, substituted into (App.II.17), yields

$$(-1) M \tilde{M} = M^N + c_1 M^{N-1} + \cdots + c_{N-1} M, \quad \text{(App.II.19)}$$

where $\tilde{M}$ is an $N \times N$ matrix that consists of the cofactor of M [see (3.84)]. Since $\det M \neq 0$, by (3.80), we obtain from (App.II.19)

$$\tilde{M} = (-1)^{N-1}(M^{N-1} + c_1 M^{N-2} + \cdots + c_{N-1} I). \quad \text{(App.II.20)}$$

It follows from (App.II.3) and (App.II.20) that

$$\sum_{\substack{j,k=1\\(j\neq k)}}^{N} (\tilde{M})_{jk}\left(\sum_{m=1}^{n} a_j^{n-m}a_k^{m-1}\right)$$
$$= (-1)^{N-1}\sum_{s=0}^{N-1} c_{N-s-1}\sum_{\substack{j,k=1\\(j\neq k)}}^{N} (M^s)_{jk}\left(\sum_{m=1}^{n} a_j^{n-m}a_k^{m-1}\right) = 0, \qquad c_0 \equiv 1. \tag{App.II.21}$$

For $n = 1, 2$, (App.II.21) become

$$\sum_{\substack{j,k=1\\(j\neq k)}}^{N} (\tilde{M})_{jk} = 0, \tag{App.II.22}$$

$$\sum_{\substack{j,k=1\\(j\neq k)}}^{N} (\tilde{M})_{jk}(a_j + a_k) = 0, \tag{App.II.23}$$

respectively.

We shall now prove (App.II.4). For this purpose, we introduce the $N \times N$ matrices B and C as

$$B_{jk} = y_j^{k-1}, \tag{App.II.24}$$

$$C = AB, \tag{App.II.25}$$

and then calculate the (j, k) component of the matrix C to obtain

$$C_{jk} = (AB)_{jk} = \sum_{s=1}^{N} A_{js}B_{sk} = y_j^{k-1}z + y_j^{k-2} + 2\sum_{s=1}^{N}{}' y_s^{k-1}(y_j - y_s)^{-1}, \tag{App.II.26}$$

where a prime appended to a sum indicates that the singular term $s = j$ is omitted.[§] Using the identity

$$\sum_{s=1}^{N}{}' y_s^{k-1}(y_j - y_s)^{-1} = \sum_{m=1}^{k-1} y_j^{m-1}\sum_{s=1}^{N}{}' y_s^{k-m-1} - y_j^{k-1}\sum_{s=1}^{N}{}'(y_s - y_j)^{-1}, \tag{App.II.27}$$

[§] The notation $\sum'$ will be used throughout this Appendix.

(App.II.26) becomes

$$C_{jk} = y_j^{k-1}z + y_j^{k-2} - 2\sum_{m=1}^{k-1} y_j^{m-1} \sum_{s=1}^{N}{}' y_s^{k-m-1} - 2y_j^{k-1}\sum_{s=1}^{N}{}'(y_s - y_j)^{-1}. \tag{App.II.28}$$

Note that the y_j ($j = 1, 2, \ldots, N$) satisfy the system of equations [see (3.297)]

$$\sum_{k=1}^{N}{}'(y_j - y_k)^{-1} = \tfrac{1}{2}(1 - y_j^{-1}), \qquad j = 1, 2, \ldots, N. \tag{App.II.29}$$

Substituting (App.II.29) into the last term on the right-hand side of (App.II.28), we obtain

$$C_{jk} = (z + 1)y_j^{k-1} - 2\sum_{m=1}^{k-1} y_j^{m-1}P_{k-m-1} + 2(k - 1)y_j^{k-2}, \tag{App.II.30}$$

where

$$P_k \equiv \sum_{s=1}^{N} y_s^k, \qquad k = 0, 1, 2, \ldots. \tag{App.II.31}$$

Then from (App.II.30)

$$\det C = \begin{vmatrix} z+1 & (z+1)y_1 - 2P_0 + 2 & (z+1)y_1^2 + (-2P_0 + 4)y_1 - 2P_1 & \cdots \\ z+1 & (z+1)y_2 - 2P_0 + 2 & (z+1)y_2^2 + (-2P_0 + 4)y_2 - 2P_1 & \cdots \\ \vdots & \vdots & \vdots & \\ z+1 & (z+1)y_N - 2P_0 + 2 & (z+1)y_N^2 + (-2P_0 + 4)y_N - 2P_1 & \cdots \end{vmatrix}$$

$$= (z+1)^2 \begin{vmatrix} 1 & y_1 & (z+1)y_1^2 + (-2P_0 + 4)y_1 - 2P_1 & \cdots \\ 1 & y_2 & (z+1)y_2^2 + (-2P_0 + 4)y_2 - 2P_1 & \cdots \\ \vdots & \vdots & \vdots & \\ 1 & y_N & (z+1)y_N^2 + (-2P_0 + 4)y_N - 2P_1 & \cdots \end{vmatrix}. \tag{App.II.32}$$

Where we have used the following properties of the determinant:

$$\begin{vmatrix} \lambda C_{11} & C_{12} & \cdots & C_{1N} \\ \lambda C_{21} & C_{22} & \cdots & C_{2N} \\ \vdots & \vdots & \ddots & \vdots \\ \lambda C_{N1} & C_{N2} & \cdots & C_{NN} \end{vmatrix} = \lambda \begin{vmatrix} C_{11} & C_{12} & \cdots & C_{1N} \\ C_{21} & C_{22} & \cdots & C_{2N} \\ \vdots & \vdots & \ddots & \vdots \\ C_{N1} & C_{N2} & \cdots & C_{NN} \end{vmatrix} \tag{App.II.33}$$

and

$$\begin{vmatrix} C_{11} & \lambda C_{11} + \mu C_{12} & \cdots & C^{1}{}_{N} \\ C_{21} & \lambda C_{21} + \mu C_{22} & \cdots & C_{2N} \\ \vdots & \vdots & \ddots & \vdots \\ C_{N1} & \lambda C_{N1} + \mu C_{N2} & \cdots & C_{NN} \end{vmatrix} = \mu \begin{vmatrix} C_{11} & C_{12} & \cdots & C_{1N} \\ C_{21} & C_{22} & \cdots & C_{2N} \\ \vdots & \vdots & \ddots & \vdots \\ C_{N1} & C_{N2} & \cdots & C_{NN} \end{vmatrix}. \tag{App.II.34}$$

Repeating this procedure, we finally arrive at the formula

$$\det C = (z+1)^N \begin{vmatrix} 1 & y_1 & \cdots & y_1^{N-1} \\ 1 & y_2 & \cdots & y_2^{N-1} \\ \vdots & \vdots & \ddots & \vdots \\ 1 & y_N & \cdots & y_N^{N-1} \end{vmatrix} = (z+1)^N \det B. \tag{App.II.35}$$

On the other hand, it follows from (App.II.25) that

$$\det C = \det AB = \det A \det B, \tag{App.II.36}$$

and from (App.II.24) that

$$\det B = \prod_{1 \le k < j \le N} (y_j - y_k) \neq 0 \tag{App.II.37}$$

by (3.296). Equating (App.II.35) and (App.II.36) yields

$$[\det A - (z+1)^N] \det B = 0, \tag{App.II.38}$$

and by (App.II.37), (App.II.38) becomes

$$\det A - (z+1)^N = 0, \tag{App.II.39}$$

which proves (App.II.4).

A technique used in obtaining (App.II.4) has an application to the algebra concerning eigenvalues of certain matrices related to the zeros of classical polynomials. The method is very interesting since it is purely algebraic. Let y_j $(j = 1, 2, \ldots, N)$ be N zeros of a certain polynomial of order N, and consider the $N \times N$ matrices $A^{(p)}$ and $C^{(p)}$ given by

$$A_{jk}^{(p)} = \left[z - \sum_{s=1}^{N}{}'(y_j - y_s)^{-p} \right] \delta_{jk} + (1 - \delta_{jk})(y_j - y_k)^{-p}, \tag{App.II.40}$$

$$C^{(p)} = A^{(p)} B, \tag{App.II.41}$$

where B is an $N \times N$ matrix defined by (App.II.24) and p a positive integer. We first calculate the (j, k) component of the matrix $C^{(p)}$ as

$$\begin{aligned} C_{jk}^{(p)} &= \sum_{s=1}^{N} A_{js}^{(p)} B_{sk} \\ &= y_j^{k-1} z - y_j^{k-1} \sum_{s=1}^{N}{}' (y_j - y_s)^{-p} + \sum_{s=1}^{N}{}' y_s^{k-1} (y_j - y_s)^{-p}. \end{aligned} \tag{App.II.42}$$

However, the identity

$$\begin{aligned} &\sum_{s=1}^{N}{}' y_s^{k-1} (y_j - y_s)^{-p} \\ &\quad = \sum_{r=0}^{p-1} (-1)^r {}_{k-1}C_r y_j^{k-r-1} \sum_{s=1}^{N}{}' (y_j - y_s)^{-(p-r)} \\ &\quad + (-1)^p \sum_{r=0}^{k-p-1} {}_{r+p+1}C_r y_j^r P_{k-p-r-1} \\ &\quad - (-1)^p \sum_{r=0}^{k-p-1} {}_{r+p-1}C_r y_j^{k-p-1} \end{aligned} \tag{App.II.43}$$

holds with

$${}_nC_r = n!/(n-r)!r!, \tag{App.II.44}$$

and P_k is defined by (App.II.31), which may be verified by mathematical induction. Substituting (App.II.43) into (App.II.42) yields

$$\begin{aligned} C_{jk}^{(p)} &= y_j^{k-1} z + \sum_{r=1}^{p-1} (-1)^r {}_{k-1}C_r y_j^{k-r-1} \sum_{s=1}^{N}{}' (y_j - y_s)^{-(p-r)} \\ &+ (-1)^p \sum_{r=0}^{k-p-1} {}_{r+p-1}C_r y_j^r P_{k-p-r-1} \\ &- (-1)^p \sum_{r=0}^{k-p-1} {}_{r+p-1}C_r y_j^{k-p-1}. \end{aligned} \tag{App.II.45}$$

We now consider, as an example, the Hermite polynomial of order N, $H_N(y)$. In this case, y_j ($j = 1, 2, \ldots, N$) satisfy the relations [125]

$$H_N(y_j) = 0, \qquad j = 1, 2, \ldots, N, \tag{App.II.46}$$

$$\sum_{s=1}^{N}{}'(y_j - y_s)^{-1} = y_j, \tag{App.II.47}$$

$$\sum_{s=1}^{N}{}'(y_j - y_s)^{-2} = \tfrac{2}{3}(N-1) - \tfrac{1}{3}y_j^2, \tag{App.II.48}$$

$$\sum_{s=1}^{N}{}'(y_j - y_s)^{-3} = \tfrac{1}{2}y_j, \tag{App.II.49}$$

$$\sum_{s=1}^{N}{}'(y_j - y_s)^{-4} = \tfrac{1}{45}[2(N+2) - y_j^2][2(N-1) - y_j^2]. \tag{App.II.50}$$

For $p = 1$, (App.II.45) becomes

$$C_{jk}^{(1)} = y_j^{k-1}z - \sum_{r=0}^{k-2} y_j^r P_{k-p-r-1} + (k-1)y_j^{k-2}, \tag{App.II.51}$$

and it follows, by employing the procedure used in deriving (App.II.35), that

$$\det C^{(1)} = z^n \det B, \tag{App.II.52}$$

and therefore

$$\det A^{(1)}(z) = \det C^{(1)}/\det B = z^n, \tag{App.II.53}$$

since $\det B \neq 0$ by the fact that $y_j \neq y_k$ for $j \neq k$. Formula (App.II.53) implies that the eigenvalues λ of the matrix $A^{(1)}(0)$, which are given by the roots of the characteristic equation

$$\det[A^{(1)}(0) - \lambda I] = 0, \qquad I: \text{unit matrix}, \tag{App.II.54}$$

are all zero.

For $p = 2$, (App.II.45) and (App.II.47) yield

$$C_{jk}^{(2)} = [z - (k-1)]y_j^{k-1} + \sum_{r=0}^{k-3}(r+1)y_j^r P_{k-r-3} - \tfrac{1}{2}(k-1)(k-2)y_j^{k-3}. \tag{App.II.55}$$

It follows from (App.II.55) that

$$\det C^{(2)} = \prod_{k=1}^{N} [z - (k - 1)] \det B, \tag{App.II.56}$$

and therefore

$$\det A^{(2)}(z) = \prod_{k=1}^{N} [x - (k - 1)], \tag{App.II.57}$$

which implies that the eigenvalues of the matrix $A^{(2)}(0)$ are all distinct and given by

$$-(k - 1), \qquad k = 1, 2, \ldots, N. \tag{App.II.58}$$

For $p = 4$, it follows from (App.II.45) and (App.II.47)–(App.II.49) that

$$\begin{aligned} C_{jk}^{(4)} = {} & [z - \tfrac{1}{6}(k^2 - 1)]y_j^{k-1} + \tfrac{1}{6}(k - 1)(k - 2)(2N - k + 1)y_j^{k-3} \\ & - \tfrac{1}{24}(k - 1)(k - 2)(k - 3)(k - 4)y_j^{k-5} \\ & + \sum_{r=0}^{k-5} \tfrac{1}{6}(r + 1)(r + 2)(r + 3)y_j^r P_{k-r-5}. \end{aligned} \tag{App.II.59}$$

Then

$$\det C^{(4)} = \prod_{k=1}^{N} [z - \tfrac{1}{6}(k^2 - 1)] \det B, \tag{App.II.60}$$

and

$$\det A^{(4)}(z) = \prod_{k=1}^{N} [z - \tfrac{1}{6}(k^2 - 1)], \tag{App.II.61}$$

which implies that the eigenvalues of the matrix $A^{(4)}(0)$ are all distinct and given by

$$-\tfrac{1}{6}(k^2 - 1), \qquad k = 1, 2, \ldots, N. \tag{App.II.62}$$

We can also show by using (App.II.45)–(App.II.50) that, for $p = 3$ and $p \geq 5$, $\det C^{(p)}$ cannot be reduced to such simple forms as (App.II.52), (App.II.56), and (App.II.60), since $C_{jk}^{(p)}$ includes the term y_j^{k+p-2} for odd p and y_j^{k+p-5} for even p, which is higher order than y_j^{k-1} for $p = 3$ and $p \geq 5$. In general, $\sum_{s=1}^{\prime N} (y_j - y_s)^{-p}$ for $p \geq 2$ is equal to a polynomial in y_j of degree p or $p - 2$, depending on whether p is even or odd [125]. In these cases the characteristic equations must be solved by other methods. The eigenvalues of $A^{(2)}(0)$ and $A^{(4)}(0)$ can also be obtained by a quite different procedure using the residue theorem [125]. However, the method presented here is purely algebraic, and it clarifies the reason why the eigenvalues of the matrices $A^{(3)}(0)$ and $A^{(p)}(0)$ $(p \geq 5)$ cannot be obtained in a simple manner. The eigenvectors may be calculated by employing Cramer's formula, but we shall not perform this calculation here.

Appendix III

Properties of the Hilbert Transform Operator

The Hilbert transform operator is defined as

$$Hf(x) = \frac{1}{\pi} \mathrm{P} \int_{-\infty}^{\infty} \frac{f(y)}{y - x} \, dy,$$

where P denotes the principal value. The following properties are fundamental:

$$H^2 f = -f, \tag{App.III.1}$$

$$\int_{-\infty}^{\infty} Hf \, dx = 0, \tag{App.III.2}$$

$$\int_{-\infty}^{\infty} f Hg \, dx = -\int_{-\infty}^{\infty} gHf \, dx, \tag{App.III.3}$$

$$Hfg = f Hg + gHf + H[(Hf)(Hg)]. \tag{App.III.4}$$

To show these formulas, define the Fourier transforms of f and g by

$$f(x) = \int_{-\infty}^{\infty} \tilde{f}(k)e^{ikx}\,dk, \tag{App.III.5}$$

$$g(x) = \int_{-\infty}^{\infty} \tilde{g}(k)e^{ikx}\,dk, \tag{App.III.6}$$

and note the following formula, which is derived using a contour integral:

$$He^{ikx} = i\,\mathrm{sgn}(k)e^{ikx}, \tag{App.III.7}$$

where

$$\mathrm{sgn}(k) = \begin{cases} 1 & \text{for } k > 0, & \text{(App.III.8a)} \\ 0 & \text{for } k = 0, & \text{(App.III.8b)} \\ -1 & \text{for } k < 0. & \text{(App.III.8c)} \end{cases}$$

Then

$$Hf = \int_{-\infty}^{\infty} \tilde{f} He^{ikx}\,dk = i \int_{-\infty}^{\infty} \tilde{f}\,\mathrm{sgn}(k)e^{ikx}\,dk, \tag{App.III.9}$$

and

$$H^2 f = i \int_{-\infty}^{\infty} \tilde{f}\,\mathrm{sgn}(k)He^{ikx}\,dk = -\int_{-\infty}^{\infty} \tilde{f}[\mathrm{sgn}(k)]^2 e^{ikx}\,dk$$

$$= -\int_{-\infty}^{\infty} \tilde{f}e^{ikx}\,dk = -f,$$

since $[\mathrm{sgn}(k)]^2 = 1$ by definition. This proves (App.III.1).

From (App.III.9) and a formula

$$\int_{-\infty}^{\infty} e^{ikx}\,dx = 2\pi\delta(k), \tag{App.III.10}$$

where $\delta(k)$ is Dirac's delta function,

$$\int_{-\infty}^{\infty} Hf\,dx = i \int_{-\infty}^{\infty} dk\,\tilde{f}\,\mathrm{sgn}(k) \int_{-\infty}^{\infty} e^{ikx}\,dx$$

$$= 2\pi i \int_{-\infty}^{\infty} \tilde{f}\,\mathrm{sgn}(k)\delta(k)\,dk = 0,$$

since

$$\mathrm{sgn}(k)\delta(k) = 0 \tag{App.III.11}$$

by (App.III.8). This proves (App.III.2).

Now consider

$$\int_{-\infty}^{\infty} fHg\,dx = \int_{-\infty}^{\infty} dx \int_{-\infty}^{\infty} dk_1 \int_{-\infty}^{\infty} dk_2 \tilde{f}(k_1)\tilde{g}(k_2)e^{ik_1x}He^{ik_2x}$$

$$= i\int_{-\infty}^{\infty} dx \int_{-\infty}^{\infty} dk_1 \int_{-\infty}^{\infty} dk_2 \tilde{f}(k_1)\tilde{g}(k_2)e^{i(k_1+k_2)x}\,\mathrm{sgn}(k_2)$$

$$= 2\pi i \int_{-\infty}^{\infty} dk_1 \int_{-\infty}^{\infty} dk_2 \tilde{f}(k_1)\tilde{g}(k_2)\delta(k_1 + k_2)\,\mathrm{sgn}(k_2)$$

$$= -2\pi i \int_{-\infty}^{\infty} \tilde{f}(k_1)\tilde{g}(-k_1)\,\mathrm{sgn}(k_1)\,dk_1, \qquad \text{(App.III.12)}$$

and similarly

$$\int_{-\infty}^{\infty} gHf\,dx = -2\pi i \int_{-\infty}^{\infty} \tilde{g}(k_1)\tilde{f}(-k_1)\,\mathrm{sgn}(k_1)\,dk_1$$

$$= 2\pi i \int_{-\infty}^{\infty} \tilde{f}(k_1)\tilde{g}(-k_1)\,\mathrm{sgn}(k_1)\,dk_1, \qquad \text{(App.III.13)}$$

where, in passing to the second equality of (App.III.13), the integration variable k_1 has been replaced by $-k_1$ and a formula $\mathrm{sgn}(-k_1) = -\mathrm{sgn}(k_1)$ has been used. Comparing (App.III.12) and (App.III.13), we obtain (App.III.3).

Formula (App.III.4) is proved as follows: Consider

$$Hfg = \int_{-\infty}^{\infty} dk_1 \int_{-\infty}^{\infty} dk_2 \tilde{f}(k_1)\tilde{g}(k_2)He^{i(k_1+k_2)}$$

$$= i\int_{-\infty}^{\infty} dk_1 \int_{-\infty}^{\infty} dk_2 \tilde{f}(k_1)\tilde{g}(k_2)\,\mathrm{sgn}(k_1 + k_2). \qquad \text{(App.III.14)}$$

On the other hand,

$$fHg + gHf + H[(Hf)(Hg)]$$

$$= \int_{-\infty}^{\infty} dk_1 \int_{-\infty}^{\infty} dk_2 \tilde{f}(k_1)\tilde{g}(k_2)\{e^{ik_1x}He^{ik_2x} + e^{ik_2x}He^{ik_1x}$$

$$+ H[(He^{ik_1x})(He^{ik_2x})]\}$$

$$= i\int_{-\infty}^{\infty} dk_1 \int_{-\infty}^{\infty} dk_2 \tilde{f}(k_1)\tilde{g}(k_2)[\mathrm{sgn}(k_2) + \mathrm{sgn}(k_1)$$

$$- \mathrm{sgn}(k_1)\,\mathrm{sgn}(k_2)\,\mathrm{sgn}(k_1 + k_2)]e^{i(k_1+k_2)x}. \qquad \text{(App.III.15)}$$

Therefore

$$\begin{aligned} Hfg &- \{fHg + gHf + H[(Hf)(Hg)]\} \\ &= i \int_{-\infty}^{\infty} dk_1 \int_{-\infty}^{\infty} dk_2 \tilde{f}(k_1)\tilde{g}(k_2)\{[1 + \operatorname{sgn}(k_1)\operatorname{sgn}(k_2)] \\ &\quad \times \operatorname{sgn}(k_1 + k_2) - \operatorname{sgn}(k_1) - \operatorname{sgn}(k_2)\}e^{i(k_1+k_2)x}. \quad \text{(App.III.16)} \end{aligned}$$

However, it is clear from the definition of function $\operatorname{sgn}(k)$ that

$$[1 + \operatorname{sgn}(k_1)\operatorname{sgn}(k_2)]\operatorname{sgn}(k_1 + k_2) = \operatorname{sgn}(k_1) + \operatorname{sgn}(k_2). \quad \text{(App.III.17)}$$

which, substituted into (App.III.16), gives (App.III.4).

It follows as a consequence of (App.III.1) and (App.III.3) that

$$\int_{-\infty}^{\infty} fHf\,dx = 0, \quad \text{(App.III.18)}$$

$$\int_{-\infty}^{\infty} (Hf)(Hg)\,dx = \int_{-\infty}^{\infty} fg\,dx. \quad \text{(App.III.19)}$$

We define the projection operators $P_{\pm}$ by

$$P_{+} = \tfrac{1}{2}(1 + iH), \quad \text{(App.III.20)}$$

$$P_{-} = \tfrac{1}{2}(1 - iH). \quad \text{(App.III.21)}$$

The following formulas are then verified using (App.III.1)–(App.III.3):

$$(P_{\pm})^2 = P_{\pm}, \quad \text{(App.III.22)}$$

$$P_{+}P_{-} = P_{-}P_{+} = 0, \quad \text{(App.III.23)}$$

$$\int_{-\infty}^{\infty} (P_{\pm} f)(P_{\mp} g)\,dx = \int_{-\infty}^{\infty} f(P_{\mp}P_{\pm})g\,dx = 0. \quad \text{(App.III.24)}$$

The following useful formula is derived using properties (App.III.1)–(App.III.4):

$$\int_{-\infty}^{\infty} dx \int_{-\infty}^{x} fHf\,dy = 0, \quad \text{(App.III.25)}$$

where the integration with respect to x should be interpreted as a principal value. It follows from (App.III.4) with $g = f$ that

$$fHf = \tfrac{1}{2}H[f^2 - (Hf)^2] = HF, \qquad \text{(App.III.26)}$$

where

$$F = \tfrac{1}{2}[f^2 - (Hf)^2]. \qquad \text{(App.III.27)}$$

It should be noted from (App.III.1) and (App.III.3) that

$$\int_{-\infty}^{\infty} F\,dx = 0. \qquad \text{(App.III.28)}$$

We denote the Fourier transform of F as $\tilde{F}$, that is,

$$F(y) = \int_{-\infty}^{\infty} \tilde{F}(k)e^{iky}\,dk, \qquad \text{(App.III.29)}$$

$$\tilde{F}(k) = 1/2\pi \int_{-\infty}^{\infty} F(y)e^{-iky}\,dy. \qquad \text{(App.III.30)}$$

From (App.III.28) and (App.III.30) we derive the relation

$$\tilde{F}(0) = 1/2\pi \int_{-\infty}^{\infty} F(y)\,dy = 0. \qquad \text{(App.III.31)}$$

Then

$$\begin{aligned} J &\equiv \int_{-\infty}^{\infty} dx \int_{-\infty}^{x} fHf\,dy \\ &= \lim_{n\to\infty} \int_{-n}^{n} dx \int_{-\infty}^{x} HF(y)\,dy \\ &= \lim_{n\to\infty} i \int_{-n}^{n} dx \int_{-\infty}^{\infty} dk\,\tilde{F}(k)\,\mathrm{sgn}(k) \int_{-\infty}^{x} e^{iky}\,dy \\ &= \lim_{n\to\infty} i \int_{-n}^{n} dx \int_{-\infty}^{\infty} dk\,\tilde{F}(k)\,\mathrm{sgn}(k)[e^{ikx}/ik + \pi\delta(k)], \qquad \text{(App.III.32)} \end{aligned}$$

where we have used (App.III.7) and a formula

$$\lim_{y \to -\infty} e^{iky}/ik = -\pi\delta(k). \qquad \text{(App.III.33)}$$

By considering (App.III.10), J is estimated as

$$J = \lim_{k \to 0} 2\pi \tilde{F}(k)/k \int_{-\infty}^{\infty} \mathrm{sgn}(k)\delta(k)\, dk$$

$$+ \lim_{n \to \infty} 2\pi i n \tilde{F}(0) \int_{-\infty}^{\infty} \mathrm{sgn}(k)\delta(k)\, dk = 0, \qquad \text{(App.III.34)}$$

since $\lim_{k \to 0} \tilde{F}(k)/k$ is finite by (App.III.31) and (App.III.11). This completes the proof of (App.III.25).

Appendix IV

Proof of (3.274)

Define functions $F(a)$ and $g(k)$ by

$$\bar{F}(a) = \begin{cases} F(a), & a > 0, \qquad \text{(App.IV.1a)} \\ 0, & a < 0, \qquad \text{(App.IV.1b)} \end{cases}$$

$$\bar{F}(a) = \int_{-\infty}^{\infty} g(k) e^{ika}\, dk. \qquad \text{(App.IV.2)}$$

Introducing (App.IV.1) and (App.IV.2) into (3.273) yields

$$\pi\beta \int_{-\infty}^{\infty} (a/4)^{n-1} \bar{F}(a)\, da = (u_0^n l/n) \int_{-\infty}^{\infty} \phi^n(\xi)\, d\xi. \qquad \text{(App.IV.3)}$$

On the other hand, it follows from (App.IV.2) that

$$\begin{aligned}\int_{-\infty}^{\infty} (a/4)^{n-1}\bar{F}(a)\, da &= \int_{-\infty}^{\infty} dk\, g(k) \int_{-\infty}^{\infty} (a/4)^{n-1} e^{ika}\, da \\ &= \int_{-\infty}^{\infty} dk\, g(k)(4i)^{-(n-1)}\, \partial^{n-1}/\partial k^{n-1} \int_{-\infty}^{\infty} e^{ika}\, da \\ &= 2\pi(i/4)^{n-1} \int_{-\infty}^{\infty} g^{(n-1)}(k)\delta(k)\, dk \\ &= 2\pi(i/4)^{n-1} g^{(n-1)}(0), \qquad \text{(App.IV.4)}\end{aligned}$$

where

$$g^{(n-1)}(k) = \partial^{n-1} g(k)/\partial k^{n-1}. \qquad \text{(App.IV.5)}$$

Substituting (App.IV.4) into (App.IV.3) gives

$$g^{(n-1)}(0) = \frac{(4u_0)^n l}{8\pi^2 \beta i^{n-1}} \frac{1}{n} \int_{-\infty}^{\infty} \phi^n(\xi)\, d\xi. \qquad \text{(App.IV.6)}$$

Thus

$$\begin{aligned}g(k) &= \sum_{n=0}^{\infty} \frac{g^{(n)}(0)}{n!} k^n \\ &= \frac{l}{8\pi^2 \beta k} \int_{-\infty}^{\infty} \sum_{n=0}^{\infty} \frac{(4u_0 k\phi)^{n+1}}{i^n (n+1)!}\, d\xi \\ &= \frac{il}{8\pi^2 \beta k} \int_{-\infty}^{\infty} (e^{-4iu_0 k\phi} - 1)\, d\xi. \qquad \text{(App.IV.7)}\end{aligned}$$

Introducing (App.IV.7) into (App.IV.2) yields

$$\begin{aligned}\bar{F}(a) &= \frac{il}{8\pi^2 \beta} \int_{-\infty}^{\infty} d\xi \int_{-\infty}^{\infty} \frac{e^{-4iu_0 k\phi} - 1}{k} e^{ika}\, dk \\ &= \frac{l}{8\pi^2 \beta} \int_{-\infty}^{\infty} d\xi \int_{-\infty}^{\infty} \frac{\sin(2u_0 k\phi)}{k} e^{ik(a - 2u_0\phi)}\, dk \\ &= \frac{l}{4\pi^2 \beta} \int_{-\infty}^{\infty} \pi h(2u_0\phi - a; -2u_0\phi, 2u_0\phi)\, d\xi, \qquad \text{(App.IV.8)}\end{aligned}$$

where use has been made of the formula

$$\int_{-\infty}^{\infty} \frac{\sin(\alpha k)}{k} e^{ikx}\, dk = \pi h(x; -\alpha, \alpha), \qquad \alpha > 0, \quad \text{(App.IV.9)}$$

with

$$h(x; -\alpha, \alpha) = \begin{cases} 1, & |x| < \alpha, \quad \text{(App.IV.10a)} \\ 0, & |x| > \alpha. \quad \text{(App.IV.10b)} \end{cases}$$

It follows from (App.IV.8)–(App.IV.10) that

$$\bar{F}(a) = \begin{cases} \sigma/4\pi u_0 \displaystyle\int_{4u_0\phi > a} d\xi, & \text{for} \quad a > 0, \quad \text{(App.IV.11a)} \\ 0, & \text{for} \quad a < 0, \quad \text{(App.IV.11b)} \end{cases}$$

which proves (3.274).

References

1. C. S. Gardner, J. M. Greene, M. D. Kruskal, and R. M. Miura. Method for solving the Korteweg–de Vries equation. *Phys. Rev. Lett.* **19**, 1095–1097 (1967).
2. P. D. Lax. Integrals of nonlinear equations of evolution and solitary waves. *Comm. Pure Appl. Math.* **21**, 467–490 (1968).
3. V. E. Zakharov and A. B. Shabat. A scheme for integrating the nonlinear equations of mathematical physics by the method of the inverse scattering problem. I. *Functional Anal. Appl.* **8**, 226–235 (1974).
4. M. J. Ablowitz, D. J. Kaup, A. C. Newell, and H. Segur. The inverse scattering transform—Fourier analysis for nonlinear problems. *Stud. Appl. Math.* **53**, 249–315 (1974).
5. R. K. Bullough and P. J. Caudrey (eds.). "Solitons," vol. 17 in "Topics in Current Physics." Springer, Berlin, 1980.
6. G. L. Lamb, Jr. "Elements of Soliton Theory." Wiley, New York, 1980.
7. M. J. Ablowitz and H. Segur. "Solitons and the Inverse Scattering Transform." SIAM, Philadelphia, 1981.
8. F. Calogero and A. Degasperis. "Spectral Transform and Solitons I," vol. 13 in

Studies in Mathematics and Applications." North-Holland Publ., Amsterdam, 1982.
9. R. Hirota. Exact solution of the Korteweg–de Vries equation for multiple collisions of solitons. *Phys. Rev. Lett.* **27**, 1192–1194 (1971).
10. R. Hirota and J. Satsuma. A variety of nonlinear network equations generated from the Bäcklund transformation for the Toda lattice. *Progr. Theoret. Phys. Suppl.* **59**, 64–100 (1976).
11. R. M. Miura (ed.), "Bäcklund transformations," Vol. 515 in "Lecture Notes in Mathematics," pp. 40–68. Springer, Berlin, 1967.
12. C. Rogers and W. F. Shadwick. "Bäcklund Transformations and Their Applications," vol. 161 in "Mathematics in Science and Engineering." Academic Press, New York, 1982.
13. R. Hirota. A new form of Bäcklund transformations and its relation to the inverse scattering problem. *Progr. Theoret. Phys.* **52**, 1498–1512 (1974).
14. T. B. Benjamin. Internal waves of finite amplitudes and permanent form. *J. Fluid Mech.* **29**, 241–270 (1966).
15. T. B. Benjamin. Internal waves of permanent form in fluids of great depth. *J. Fluid Mech.* **29**, 559–592 (1967).
16. R. E. Davis and A. Acrivos. Solitary internal waves in deep water. *J. Fluid Mech.* **29**, 593–607 (1967).
17. H. Ono. Algebraic solitary waves in stratified fluids. *J. Phys. Soc. Japan* **39**, 1082–1091 (1975).
18. H. Ono. Algebraic Rossby wave soliton. *J. Phys. Soc. Japan* **50**, 2757–2761 (1981).
19. Y. Matsuno. *N*-soliton and *N*-periodic wave solutions of the higher-order Benjamin–Ono equation. *J. Phys. Soc. Japan* **47**, 1745–1746 (1979).
20. Y. Matsuno. Solutions of the higher-order Benjamin–Ono equation. *J. Phys. Soc. Japan* **48**, 1024–1028 (1980).
21. Y. Matsuno. Bilinearization of nonlinear evolution equations IV. Higher-order Benjamin–Ono equations. *J. Phys. Soc. Japan* **49**, 1584–1592 (1980).
22. Y. Matsuno. Bilinearization of nonlinear evolution equations. *J. Phys. Soc. Japan* **48**, 2138–2143 (1980).
23. R. I. Joseph. Solitary waves in a finite depth fluid. *J. Phys. A* **10**, L225–L227 (1977).
24. R. I. Joseph and R. Egri. Multi-soliton solutions in a finite depth fluid. *J. Phys. A* L97–L102 (1978).
25. T. Kubota, D. R. S. Ko, and L. D. Dobbs. Weakly-nonlinear, long internal gravity waves in stratified fluids of finite depth. *J. Hydronautics* 157–165 (1978).
26. H. Ono. Nonlinear Rossby waves in a channel of finite width. *J. Phys. Soc. Japan* **51**, 2318–2325 (1982).
27. Y. Matsuno. Exact multi-soliton solution for nonlinear waves in a stratified fluid of finite depth. *Phys. Lett. A* **74**, 233–235 (1979).
28. A. Nakamura and Y. Matsuno. Exact one- and two-periodic wave solution of fluids of finite depth. *J. Phys. Soc. Japan* **48**, 653–657 (1980).
29. Y. Matsuno. *N*-soliton solution of the higher order wave equation for a fluid of finite depth. *J. Phys. Soc. Japan* **48**, 663–668 (1980).
30. Y. Matsuno. Bilinearization of nonlinear evolution equations II. Higher-order modified Korteweg–de Vries equations. *J. Phys. Soc. Japan* **49**, 787–794 (1980).
31. Y. Matsuno. Bilinearization of nonlinear evolution equations III. Bäcklund transformations of higher-order Korteweg–de Vries equations. *J. Phys. Soc. Japan* **49** 795–801 (1980).

32. A. Nakamura. N-periodic wave and N-soliton solutions of the modified Benjamin–Ono equation. *J. Phys. Soc. Japan* **47**, 2045–2046 (1979).
33. M. Tanaka. Nonlinear self-modulation of interfacial waves. *J. Phys. Soc. Japan* **51** 2016–2023.
34. M. Tanaka. Nonlinear self-modulation problem of the Benjamin–Ono equation. *J. Phys. Soc. Japan* **51**, 2686–2692 (1982).
35. D. J. Korteweg and G. de Vries. On the change of form of long waves advancing in a rectangular canal, and on a new type of long stationary waves. *Philos. Mag.* **39**, 422–443 (1895).
36. A. Jeffrey and T. Kakutani. Weak nonlinear dispersive waves: A discussion centered around the Korteweg–de Vries equation. *SIAM Review* **14**, 582–643 (1972).
37. A. C. Scott, F. Y. F. Chu, and D. W. McLaughlin. The soliton—A new concept in applied science. *Proc. IEEE* **61**, 1443–1483 (1973).
38. R. M. Miura. The Korteweg–de Vries equation; A survey of results. *SIAM Rev.* **18**, 412–459 (1976).
39. R. Hirota. Bilinearization of soliton equations. *J. Phys. Soc. Japan* **51**, 323–331 (1982).
40. K. Sawada and T. Kotera. A method for finding N-soliton solutions of the K.d.V. equation and K.d.V.-like equation. *Progr. Theoret. Phys.* **51**, 1355–1367 (1974).
41. M. Wadati and M. Toda. The exact N-soliton solution of the Korteweg–de Vries equation. *J. Phys. Soc. Japan* **32**, 1403–1411 (1972).
42. S. Tanaka. On the N-tuple wave solutions of the Korteweg–de Vries equation. *Publ. Res. Inst. Math. Sci.* **8**, 419–427 (1972/3).
43. C. S. Gardner, J. M. Greene, M. D. Kruskal, and R. M. Miura. Korteweg–de Vries equation and generalizations. VI. Methods for exact solution. *Comm. Pure Appl. Math.* **27**, 97–133 (1974).
44. R. R. Rosales. Exact solutions of some nonlinear evolution equations. *Stud. Appl. Math.* **59**, 117–151 (1978).
45. S. Oishi. Relationship between Hirota's method and the inverse spectral method—The Korteweg–de Vries equation's case. *J. Phys. Soc. Japan* **47**, 1037–1038 (1979).
46. S. Oishi. A method of constructing generalized soliton solutions for certain bilinear soliton equations. *J. Phys. Soc. Japan* **47**, 1341–1346 (1979).
47. S. Oishi. A method of analyzing soliton equations by bilinearization. *J. Phys. Soc. Japan* **48**, 639–646 (1980).
48. A. Nakamura. A direct method of calculating periodic wave solutions to nonlinear evolution equations. I. Exact two-periodic wave solution. *J. Phys. Soc. Japan* **47**, 1701–1705 (1979).
49. A. Nakamura. A direct method of calculating periodic wave solutions to nonlinear evolution equations. II. Exact one- and two-periodic wave solution of the coupled bilinear equations. *J. Phys. Soc. Japan* **48**, 1365–1370 (1980).
50. H. D. Wahlquist and F. B. Estabrook. Bäcklund transformations for solutions of the Korteweg–de Vries equation. *Phys. Rev. Lett.* **31**, 1386–1390 (1973).
51. R. Hirota and J. Satsuma. A simple structure of superposition formula of the Bäcklund transformation. *J. Phys. Soc. Japan* **45**, 1741–1750 (1978).
52. A. Nakamura and R. Hirota. Second modified KdV equation and its exact multi-soliton solution. *J. Phys. Soc. Japan* **48**, 1365–1370 (1980).
53. A. Nakamura. Chain of Bäcklund transformation for the KdV equation. *J. Math. Phys.* **22**, 1608–1613 (1981).
54. I. M. Gel'fand and L. A. Dikii. Asymptotic behavior of the resolvent of Sturm–

Liouville equations and the algebra of the Korteweg–de Vries equations. *Russian Math. Surveys* **30**, 77–133 (1975).
55. R. Hirota. Exact solution of the modified Korteweg–de Vries equation for multiple collisions of solitons. *J. Phys. Soc. Japan* **33**, 1456–1458 (1972).
56. R. Hirota. Exact solution of the Sine–Gordon equation for multiple collisions of solitons. *J. Phys. Soc. Japan* **33**, 1459–1463 (1972).
57. R. Hirota. Exact envelope-soliton solutions of a nonlinear wave equation. *J. Math. Phys.* **14**, 805–809 (1973).
58. R. Hirota. Exact N-soliton solutions of the wave equation of long waves in shallow-water and in nonlinear lattice, *J. Math. Phys.* **14**, 810–814 (1973).
59. R. Hirota and J. Satsuma. N-soliton solutions of model equations for shallow water waves. *J. Phys. Soc. Japan* **40**, 611–612 (1976).
60. R. Hirota. Exact solutions to the equation describing cylindrical solitons." *Phys. Lett. A* **71**, 393–394 (1979).
61. A. Nakamura and H. H. Chen. Soliton solution of the cylindrical KdV equation. *J. Phys. Soc. Japan* **50**, 711–718 (1981).
62. A. Nakamura and H. H. Chen. Multi-soliton solution of a derivative nonlinear Schrödinger equation. *J. Phys. Soc. Japan* **49**, 813–816 (1980).
63. Y. Matsuno. Exact multi-soliton solution of the Benjamin–Ono equation. *J. Phys. A* **12**, 619–612.
64. J. Satsuma and Y. Ishimori. Periodic wave and rational soliton solutions of the Benjamin–Ono equation. *J. Phys. Soc. Japan* **46**, 681–687 (1979).
65. H. H. Chen and Y. C. Lee. Internal wave solitons of fluids with finite depth. *Phys. Rev. Lett.* **43**, 264–266 (1979).
66. R. Hirota. Exact three-soliton solution of the two-dimensional Sine–Gordon equation. *J. Phys. Soc. Japan* **35**, 1566 (1973).
67. J. Satsuma. N-soliton solution of the two-dimensional Korteweg–de Vries equation. *J. Phys. Soc. Japan* **40**, 286–290 (1976).
68. A. Nakamura. One dimensionally aligned decay mode solutions of the two dimensional nonlinear Schrödinger equation. *J. Phys. Soc. Japan* **50**, 2469–2470 (1981).
69. R. Hirota. Exact N-soliton solutions of a nonlinear lumped network equation. *J. Phys. Soc. Japan* **35**, 286–288 (1973).
70. M. Toda. "Theory of Nonlinear Lattices," vol. 20 in Springer Series in Solid-State Sciences." Springer, Berlin, 1981.
71. R. Hirota. Exact N-soliton solutions of nonlinear self-dual network equations. *J. Phys. Soc. Japan* **35**, 289–294 (1973).
72. R. Hirota and J. Satsuma. N-soliton solutions of nonlinear network equations describing a Volterra system. *J. Phys. Soc. Japan* **40**, 891–900 (1976).
73. R. Hirota. Nonlinear partial difference equations. I. A. difference analogue of the Korteweg–de Vries equation. *J. Phys. Soc. Japan* **43**, 1424–1433 (1977).
74. R. Hirota. Nonlinear partial difference equations. II. Discrete-time Toda equation. *J. Phys. Soc. Japan* **43**, 2074–2078 (1977).
75. R. Hirota. Nonlinear partial difference equations. III. Discrete Sine–Gordon equation. *J. Phys. Soc. Japan* **43**, 2079–2086 (1977).
76. R. Hirota. Nonlinear partial difference equations. IV. Bäcklund transformation for the discrete-time Toda equation. *J. Phys. Soc. Japan* **45**, 321–332 (1978).

77. R. Hirota. Nonlinear partial difference equations. V. Nonlinear equations reducible to linear equations. *J. Phys. Soc. Japan* **46**, 312–319 (1979).
78. R. Hirota. Discrete analogue of a generalized Toda equation. *J. Phys. Soc. Japan* **50**, 3785–3791 (1981).
79. R. Hirota and J. Satsuma. Nonlinear evolution equations generated from the Bäcklund transformation for the Toda lattice. *J. Phys. Soc. Japan* **40**, 891–900 (1976).
80. R. Hirota and J. Satsuma. Nonlinear evolution equations generated from the Bäcklund transformation for the Boussinesq equation. *Progr. Theoret. Phys.* **57**, 797–807 (1977).
81. A. Nakamura. Exact N-soliton solution of the modified finite depth fluid equation. *J. Phys. Soc. Japan* **47**, 2043–2044 (1979).
82. A. Nakamura. Exact multi-soliton solution of the modified Sine–Gordon equation. *J. Phys. Soc. Japan* **49**, 1167–1170 (1980).
83. S. Oishi. The Korteweg–de Vries equation under slowly decreasing boundary condition. *J. Phys. Soc. Japan* **48**, 349–350 (1980).
84. M. J. Ablowitz and J. Satsuma. Solitons and rational solutions of nonlinear evolution equations. *J. Math. Phys.* **19**, 2180–2186 (1978).
85. A. Nakamura. Simple similarity-type multiple-decay-mode solution of the two-dimensional KdV equation. *Phys. Rev. Lett.* **46**, 751–753 (1989).
86. A. Nakamura. Decay mode solution of the two-dimensional KdV equation and the generalized Bäcklund transformation. *J. Math. Phys.* **22**, 2456–2462 (1981).
87. A. Nakamura. One dimensionally aligned decay mode solutions of the two-dimensional nonlinear Schrödinger equation. *J. Phys. Soc. Japan* **50**, 2467–2470 (1981).
88. A. Nakamura. Exact Bessel type solution of the two-dimensional Toda lattice equation. *J. Phys. Soc. Japan* **52**, 380–387 (1983).
89. Y. Matsuno. On the Benjamin–Ono equation—Method for exact solution. *J. Phys. Soc. Japan* **51**, 3734–3739 (1982).
90. K. M. Case. The N-soliton solution of the Benjamin–Ono equation. *Proc. Nat. Acad. Sci. U.S.A.* **75**, 3562–3563 (1978).
91. H. H. Chen, Y. C. Lee, and N. R. Pereira. Algebraic internal wave solitons and the integrable Calogero–Moser–Sutherland N-body problem. *Phys. Fluids* **22**, 187–188 (1979).
92. Y. Matsuno. Algebra related to the N soliton solution of the Benjamin–Ono equation. *J. Phys. Soc. Japan* **51**, 2719–2720 (1982).
93. M. D. Kruskal. The Korteweg–de Vries equation and related evolution equations. *Lect. Appl. Math., Am. Math. Soc.* **15**, 61–83 (1974).
94. W. R. Thickstun. A system of particles equivalent to solitons. *J. Math. Anal. Appl.* **55**, 335–346 (1976).
95. H. Airault, H. P. Mckean, and J. Moser. On a class of polynomials connected with the Korteweg–de Vries equation. *Comm. Math. Phys.* **61**, 1–30 (1978).
96. D. V. Choodnovsky and G. V. Choodnovsky. Pole expansion of nonlinear differential equations. *Nuovo Ciminto B* **40**, 339–353 (1977).
97. F. Calogero. Motion of poles and zeros of special solutions of nonlinear and linear partial differential equations and related "solvable" many-body problems. *Nuovo Cimento B* **43**, 177–241 (1978).
98. K. M. Case. Properties of the Benjamin–Ono equation. *J. Math. Phys.* **20**, 972–977 (1977).

99. K. M. Case. Meromorphic solutions of the Benjamin–Ono equation. *Phys. A* **96,** 173–182 (1979).
100. J. Moser. Three integrable Hamiltonian systems connected with isospectral deformations. *Adv. Math.* **16,** 197–220 (1975).
101. K. M. Case. Benjamin–Ono-related equations and their solutions. *Proc. Nat. Acad. Sci. U.S.A.* **76,** 173–182 (1979).
102. K. M. Case. The Benjamin–Ono and related equations [1]. *Phys. D* **3,** 185–192 (1981).
103. A. Nakamura. Bäcklund transform and conservation laws of the Benjamin–Ono equation. *J. Phys. Soc. Japan* **47,** 1335–1340 (1979).
104. T. L. Bock and M. D. Kruskal. A two-parameter Miura transformation of the Benjamin–Ono equation. *Phys. Lett. A* **74,** 173–176 (1979).
105. R. M. Miura. Korteweg–de Vries equation and generalizations. I. A remarkable explicit nonlinear transformation. *J. Math. Phys.* **9,** 1202–1204 (1968).
106. Y. Matsuno. Recurrence formula and conserved quantity of the Benjamin–Ono equation. *J. Phys. Soc. Japan* **52,** 2955–2958 (1983).
107. A. S. Fokas and B. Fuchssteiner. The hierarchy of the Benjamin–Ono equation. *Phys. Lett. A* **86,** 341–345 (1981).
108. H. H. Chen, Y. C. Lee, and J.-E. Lin. On a new hierarchy of symmetries for the Benjamin–Ono equation. *Phys. Lett. A* **91,** 381–383 (1982).
109. L. J. F. Broer and H. M. M. Ten Eikelder. Constants of the motion for the Benjamin–Ono and related equations. *Phys. Lett. A* **92,** 56–58 (1982).
110. Y. Matsuno. Number density function of Benjamin–Ono solitons. *Phys. Lett. A* **87,** 15–17 (1981).
111. Y. Matsuno. Asymptotic properties of the Benjamin–Ono equation. *J. Phys. Soc. Japan* **51,** 667–674 (1982).
112. Y. Matsuno. Soliton and algebraic equation. *J. Phys. Soc. Japan* **51,** 3375–3380 (1982).
113. H. Hochstadt. "The Functions of Mathematical Physics." Wiley, New York, 1971.
114. Y. Matsuno. Interaction of the Benjamin–Ono solitons. *J. Phys. A* **13,** 1519–1536 (1980).
115. J. Satsuma, M. J. Ablowitz, and Y. Kodama. On an internal wave equation describing a stratified fluid with finite depth. *Phys. Lett. A* **73,** 283–286 (1979).
116. M. Adler and J. Moser. On a class of polynomials connected with the Korteweg–de Vries equation, *Comm. Math. Phys.* **61,** 1–30 (1978).
117. Y. Matsuno. The Bäcklund transformations of the higher-order Korteweg–de Vries equation. *Phys. Lett. A* **77,** 100–102 (1980).
118. D. J. Kaup and A. C. Newell. An exact solution for a derivative nonlinear Schrödinger equation. *J. Math. Phys.* **19,** 798–801 (1978).
119. S. A. Maslowe and L. G. Redekopp. Long nonlinear wave in stratified shear flows. *J. Fluid Mech.* **101,** 321–348 (1980).
120. A. H. Nayfeh, "Perturbation Methods." Wiley, New York, 1973.
121. A. Jeffrey and T. Kawahara. "Asymptotic Methods in Nonlinear Wave Theory." Pitman, London, 1982.
122. M. Tanaka. Perturbations of the K–dV solitons—An approach based of the multiple time scale expansion. *J. Phys. Soc. Japan* **49,** 807–812 (1980).
123. J. P. Keener and D. W. Mclaughlin. Solitons under perturbations. *Phys. Rev. A* **16,** 777–790 (1977).

124. Y. Kodama and M. J. Ablowitz. Perturbations of solitons and solitary waves. *Stud. Appl. Math.* **64**, 225–245 (1981).
125. S. Ahmed, M. Bruschi, F. Calogero, M. A. Olshanetsky, and A. M. Perelomov. Properties of the zeros of the classical polynomials and of the Bessel functions. *Nuovo Cimento B* **49**, 173–199 (1979).

Author Index

Subject Index